Acknowledgments

The authors would like to express their appreciation to several people who reviewed the draft of this book and provided us with comments and suggestions to improve it. Our thanks go to Marlin Hacker and Don Ulrich at Morgan Products, Bob Watson at Rolls-Royce Motor Cars, and Chris Cage of ICI Pharmaceuticals for their help and assistance. We would especially like to recognize the input from John Cook of Reckitt and Colman and Jim Wood of Teradyne, Inc., which resulted in our adding some important information and assisted us in dealing with some difficult issues.

There are a number of our colleagues at the Oliver Wight Companies who gave us a tremendous amount of feedback. They include Bill Boyst, Larry Curry, Pete Foy, Walt Goddard, and Darryl Landvater. Pete kept us on the mark in our discussions of remanufacturing, and Larry made several suggestions that contributed to tightening up the text.

As anyone who has ever written a book knows, there are a few people behind the scenes who help make the book possible. Thanks go to Rachel Snyder for her help in typing the myriad drafts and keeping them all straight. Next, a tip of the hat to Ron Schultz, who made continuous improvements to our writings and really made it all come together. Finally, our appreciation goes to Dana Scannell and Jim Childs, our editors, who continually cracked the whip and kept this project moving ahead.

Finally, we would like to thank all those folks who have purchased this book and have at least read these acknowledgments. The book is ultimately for them, and its true success will be measured by how well they are able to implement the tools we discuss here.

                                             Jerry Clement
                                             Hamlin, New York

                                             Andy Coldrick
                                             Gloucester, England

                                             John Sari
                                             Pfafftown, North Carolina

Contents

CHAPTER 1
Overview — 3

CHAPTER 2
Defining the Manufacturing Process — 17

CHAPTER 3
Defining Levels in the Bills of Material and Routings — 47

CHAPTER 4
Achieving Accuracy and Completeness — 61

CHAPTER 5
Planning Scheduling and Controlling the Plant Using the Data Foundation — 73

CHAPTER 6
Modularizing the Bill of Material — 99

CHAPTER 7
New Product Introduction and Custom Manufacturing — 139

CHAPTER 8
Managing Engineering Change Control — 161

CHAPTER 9
Implementing Change — 187

APPENDICES
Appendix A
By-products and Co-products — 205

Appendix B
Preventive Maintenance — 217

Appendix C
Refurbishing, Remanufacturing, and Reconditioning — 221

Appendix D
Reprocessing — 235

Appendix E
Tool Requirements Planning — 241

Glossary — 247

Index — 271

Manufacturing
Data Structures

Chapter 1
Overview

The dinner rush has begun at the Hunt House Continental Restaurant. Waiters hurry into the kitchen to place their orders from the hungry customers out front. Cooks are busily satisfying those requests—broiling steaks, sautéing meats and fish, preparing and ladling sauces, chopping salads, decorating desserts, and assembling the plates with the proper combinations of vegetables, potatoes or rice, and garnish.

Harvey Allen, general manager of the restaurant, calmly walks through the kitchen, surveying the evening's progress. He nods his satisfaction to the head chef, Charles Milliken, then leaves the kitchen and greets a number of regular guests on his way to the maître d's stand. He checks the reservations. Business is good. But then, that is no accident. Together with his head chef, maître d', and business partners, Allen has blended together the various elements of his operation in much the same way that Chef Milliken combines the precise ingredients of fish, spices, and wine to make his special bouillabaisse.

There was a time, however, when running the Hunt House seemed an overwhelming task. Operations were truly bogged down. Orders floundered in the kitchen and then were sent back cold or undercooked. Some items, like the chef's special chicken marengo, took 25 minutes to prepare, while a dish like veal pomadoro took only 10 minutes. Putting an order of the two together was a constant problem. Then there was the frustrating predicament of ingredients for some dishes never being available. And what's eggplant Parmesan without the eggplant?

Another problem was that the restaurant was continually running out of its most popular dishes, while the waste from those that weren't as popular was staggering. It always seemed that there was too little inventory of what was wanted and too much of what wasn't. Management was always ordering the employees to reduce the restaurant's expenses and come up with ideas to cut down on the waste. They needed a way to simplify the process. That's when Harvey Allen stepped in and took control of operations.

Allen implemented a system that focused on the needs of each individual area of the restaurant, in this way serving the needs of the whole company. With this central core system, all areas of the restaurant are linked together to produce the success that the Hunt House has become.

Within the daily routine, Allen and Chef Milliken are responsible not only for planning the menu and designing the recipes for each dish, but also for ordering everything that will be needed by the restaurant. This means that they have to be able to assess what dishes on the menu are most popular—how many filet mignons versus New York steaks are served, whether they are selling more of the duck with orange sauce or the lamb curry—and which ingredients are necessary for those dishes. Once the dinner requirements are forecast, the recipes identify all those food items needed to prepare and serve those dinners. Then Allen and Milliken need to consider the various lead times necessary to acquire the various food items.

Another part of the chef's job is to decide what specials he'll be preparing for the coming week. He has to figure out what raw materials he needs, how many salads can be made from a crate of lettuce after removing the waste, which meats need to be portion controlled, how many triple chocolate tortes to prepare, and what proportions of dark chocolate, white chocolate, and milk chocolate to order. Again, this is all information gleaned from Chef Milliken's recipes.

Milliken also has to decide what new recipes might be added or previous ones modified. He needs to establish what substitutions can be made if a certain ingredient for a menu item isn't available, what ingredients—such as fresh fish—might only be available on a limited basis, and what ingredients might be seasonal.

He and his staff now carefully schedule the dishes that can be prepared ahead of time. They also oversee the general daily preparation that needs to be done to support those dishes, including which sauces can be made in advance and which need to be prepared at the time of serving. During the cooking process, Milliken has to know how long each process step takes, so that when a table orders one dish with a long preparation time and another dish with a short preparation time, both dishes arrive together on time, just as the customers ordered them. The timing for those process steps also lets him know when to reduce the wine for the bordelaise sauce, at what point to stir the melted butter into the hollandaise, or when to add the sour cream to the Stroganoff.

To make sure that each dish goes to the customer exactly the same way each time, Chef Milliken checks that his recipes are strictly followed. Of course, each cook has his or her own way of doing things, so the chef must randomly sample the work of his employees to be certain that the quality is what the customer at the restaurant has come to expect.

The Hunt House has an established clientele, but Harvey Allen and his business partners have recently begun a new ad campaign that is attracting a lot of new business. Allen must now take into account projections for those future increases, including upcoming holidays. Mother's Day, for example, is the biggest restaurant day of the year.

It is necessary to anticipate any extra labor they might need. Together with Chef Milliken, Allen must decide how many cooks to add on weekends and holidays. Allen must also work with the maître d' to figure out the number of busboys and waiters they will need, as well as devising their schedules and hours. Then there are the linen suppliers, the dinnerware and glassware suppliers, and the florist—not to mention their fluctuating price structures and delivery schedules to consider.

The Hunt House has brought together these different aspects of the restaurant by taking its lead from its chef. In effect, management has created a single company data foundation as a *recipe* for the restaurant. This data base of information contains not only a list of products and ingredients, but also a definition of the restaurant's processes. For the Hunt House to operate properly, it is necessary to maintain a complete set of data. This means that management needs to make sure that it has the right dishes and ingredients, the proper mix of ingredients to make each specific dish, the correct sequence of steps necessary to prepare the food with the correct timing, the necessary equipment for its preparation, and the cost and order quantities for purchased and manufactured materials—all the necessary elements that form a single company data foundation.

Chef Milliken organizes his responsibilities around the central data foundation. As part of this information, the chef maintains his recipes, from which he can then figure what he needs to purchase. He keeps track of his on-hand supplies. He can modify his recipes, changing either ingredients or method of preparation. He charts dish popularity, counts the meals prepared on a daily basis, and schedules every function in his kitchen down to the preparation of the last strawberry mousse parfait.

The maître d' uses this information to schedule his reservations, his seating arrangements, his staff, and his supplies. Because these decisions can also affect the chef's decisions, that information is communicated back to him. The waiters use their knowledge of the chef's combination of ingredients to describe the various dishes being offered, to whet the appetites of the restaurant's guests, and to push certain specials or dishes the restaurant would like to sell. With five different chicken entrées, four different veal dishes, and ten fresh fish dishes, all prepared to order, the waiters also need to know the various options available on the menu. In addition, the restaurant's accounting firm taps into portions of this same information and uses it for its accounting needs and projections. It can feed back cost analyses and provide sales history information.

The point is: Everybody uses this single data base of information to aid in the smooth operation of the restaurant.

The system that Allen and his partners implemented is what we will be describing throughout this book. We will show how this single company data foundation supports both the way a product is produced and the way it is sold. Manufacturing companies, just like the staff at the Hunt House, need good business practices and systems supported by solid company data foundations.

The need to effectively plan, schedule, and create a desired level of control exists in any manufacturing concern, whether the product being produced is ratatouille, a sophisticated radar system, nuts and bolts, or prescription drugs. Regardless of their product and manufacturing processes, companies often struggle with the following kinds of problems:

- Not buying or making the right items at the right time
- Missed customer deliveries
- Too much inventory
- Little improvement in productivity
- High costs
- Frustration and low morale caused by ineffective operating practices

There are companies, however, that don't suffer from these problems. Their customer service is excellent, inventories are low, productivity is up, and morale is high. A major contributor to such success has been the implementation of Manufacturing Resource Planning, or MRP II.

Manufacturing Resource Planning is a company-wide business system that defines the material and capacity resource demands of a company and then plans the supply of those resources. It is the accepted manner of maintaining resource support in all types of manufacturing environments.

Manufacturing Resource Planning basically requires three things:

1. People who understand its use
2. An accurate, complete data foundation
3. The right software

At the core of an MRP II system is a set of computer logic that asks three basic questions:

1. What are we going to make?
2. What does it take to make it?
3. What do we have?

By addressing these questions, the system then answers a fourth:

4. What do we need, and when?

This book is all about question 2—What does it take to make it? Put another way, how does a business define its data requirements to support its manufacturing processes and its customer needs?

The data foundation shown (see Figure 1-1) is typical. To understand how a company's information might be structured, let's review some traditional manufacturing practices and see their impact on the data base design.

Manufacturing Tradition

Traditional manufacturing thinking has been to reduce cost with the greatest possible efficiency and utilization of both equipment and labor. The obstacle has always been nonproductive time, including setup time. Long runs and large build quantities became the order of the day, offset

**Figure 1-1
Manufacturing Resource Planning (MRP II)**

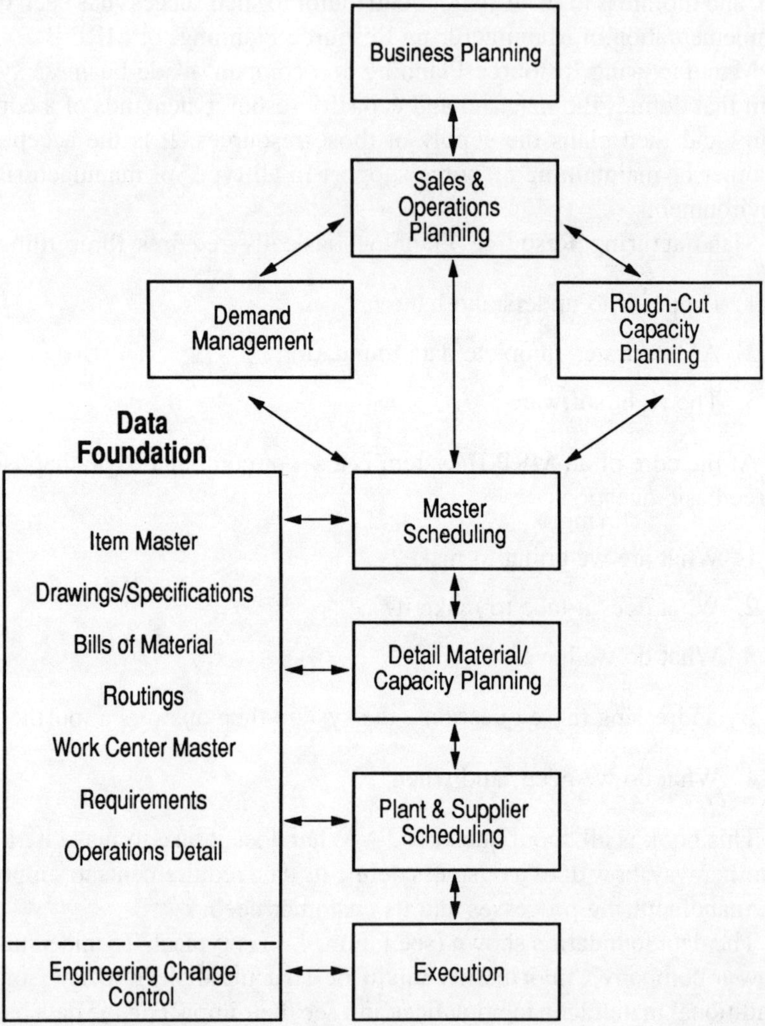

```
                        ┌───────────────────┐
                        │ Business Planning │
                        └─────────┬─────────┘
                                  ↕
                        ┌───────────────────┐
                        │     Sales &       │
                        │   Operations      │
                        │    Planning       │
                        └─────────┬─────────┘
              ↙                   ↕                   ↘
    ┌──────────────┐                          ┌──────────────┐
    │    Demand    │                          │  Rough-Cut   │
    │  Management  │                          │   Capacity   │
    │              │                          │   Planning   │
    └──────────────┘                          └──────────────┘

Data
Foundation
┌─────────────────────────────┐    ┌───────────────────┐
│                             │    │      Master       │
│      Item Master            │↔   │    Scheduling     │
│                             │    └─────────┬─────────┘
│  Drawings/Specifications    │              ↕
│                             │    ┌───────────────────┐
│     Bills of Material       │↔   │  Detail Material/ │
│                             │    │ Capacity Planning │
│        Routings             │    └─────────┬─────────┘
│                             │              ↕
│    Work Center Master       │    ┌───────────────────┐
│                             │↔   │  Plant & Supplier │
│      Requirements           │    │    Scheduling     │
│                             │    └─────────┬─────────┘
│    Operations Detail        │              ↕
│─ ─ ─ ─ ─ ─ ─ ─ ─ ─ ─ ─ ─ ─ ─│    ┌───────────────────┐
│   Engineering Change        │↔   │     Execution     │
│        Control              │    └───────────────────┘
└─────────────────────────────┘
```

only by inventory carrying costs. Economic order quantity (EOQ) formulas were used to achieve the right level of inventories by balancing the cost of more inventory against the cost to set up or purchase an item. In many manufacturing environments, this mind-set still prevails.

In such environments, product and processes did not flow. Inventory was built in large quantities and stored for later manufacturing use. Once this stock was nearly depleted, another large quantity was generated to supply future demand. This process led to a great many intermediate levels of stocking and the assignment of a great many item numbers, all of which needed to be structured into formulas, bills of material, or recipes. It was one of the reasons why deep, multilevel bill of material structures were found in manufacturing companies' data files. Large batch runs or order quantities seem to have been the solution that allowed a company to achieve greater productivity.

Another objective was to maximize the output or efficiency of each manufacturing operation as the order was processed. To do this, sequence routings were devised together with performance standards to measure the actual work performed in each work center. These step-by-step or operation-by-operation routings identified most detailed activities in the manufacturing process. Once detailed routings and standards were developed, they were used to track and monitor the variance between actual work performance and the standards. The desire to monitor work at such a detailed level led to the lengthy routings found in many manufacturing company data files.

Traditionally, then, large batches of work moved through the plant in operation sequences, work center by work center, until a stocking level was reached. At that point, the material was moved to store's inventory to await the next series of manufacturing operations. Each batch of work was controlled by a work order, and operation-level reporting was used to determine the progress and update the status of the order through the manufacturing process. Systems like detailed capacity requirements planning, shop floor control, input/output control, work-in-process (WIP) accounting, and standard and actual costing systems were used to plan and schedule operations effectively and to track and report financials accurately. Supporting all of this were large, complex data files with staffs of personnel to manage and maintain the needed information structures.

BREAKING TRADITION

These traditional objectives in manufacturing are being challenged today as manufacturing professionals are questioning the value of large inventories, multilevel manufacturing processes with long lead times, and complex data requirements. This is especially so in Just-in-Time and Total Quality Control (JIT/TQC) environments.

Walt Goddard, in his book *Just-in-Time: Surviving by Breaking Tradition*, defines Just-in-Time as "an approach to achieving excellence in a manufacturing company based on the continuing elimination of waste and consistent improvement in productivity." Waste is then defined as "those things that do not add value to the product."

In his book *Just-in-Time: Making It Happen*, Bill Sandras refers to Just-in-Time as a continuous improvement process, where the total company is discontented with the status quo. Sandras recommends a "One Less at a Time" approach that continuously reduces all wasteful activity (see Figure 1-2). The "One Less at a Time" process is all-encompassing and attacks waste at every step in the total business process (see Figure 1-3).

A cornerstone of Just-in-Time is Total Quality Control or total quality management. Just-in-Time identifies waste; Total Quality Control provides the problem-solving mechanisms to eliminate waste.

Figure 1-2
Just-in-Time/Total Quality Control (JIT/TQC)
The "One Less at a Time" Process

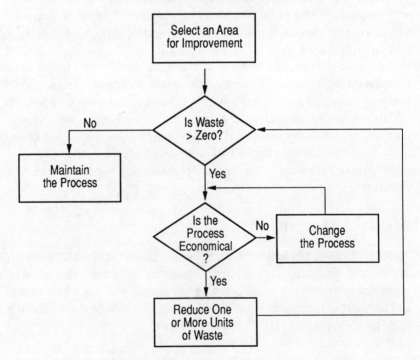

Figure 1-3
Waste in the Manufacturing Process

JIT/TQC Means Repeated Reductions . . .

In the Order Quantity
In the Safety Stock
In the Queue
In the Reject Bin
In the Setup
In the Moves on the Plant Floor
In the Number of Manufacturing Operations
In Pieces of Paper
In Complexity in the Product
In the Number of Suppliers
In Time in Order Entry
In Days of Customer Lead Time
In Warranty Claims
In Customer Returns

. . . Over and Over,
Day after Day.

If we compare JIT/TQC environments with the traditional manufacturing environments discussed earlier, we see less material stocking because of setup reduction and increased manufacturing flow and less operation-by-operation production tracking as a result of flexible labor, cellular manufacturing, and kanban scheduling. Movement in this direction has resulted in flat bills of material (fewer levels), shallow routings (fewer detailed operations), work center grouping (fewer work centers), and far fewer item numbers in the system.

THE DATA FOUNDATIONS

Manufacturing Resource Planning (MRP II) remains a cornerstone of a manufacturing company utilizing JIT/TQC philosophies. Planning, scheduling, and business control continue as important business requirements. As JIT simplifies products and processes, MRP II systems become much simpler, too. Data foundations shrink to only the information required to support a more simplified manufacturing operation.

Regardless of the manufacturing environment, MRP II systems and the concept of continuous improvement require that a company's information be identified and maintained in support of the company's planning, scheduling, and reporting. In almost all companies today, that requires computer software and related data base files. This means that items need to be identified, material content must be defined, the manufacturing operation must be described, and the resulting products and semifinished materials need to be depicted in drawings or specifications. The data files must be designed and organized to reflect the chosen manufacturing process.

Data foundations are also influenced by the product being produced and by the way a company stages or positions products for its customers. Some companies sell off the shelf, while some produce to lower levels—process batch or subassembly level—and then pack, finish, or assemble to customer requirements. Other companies begin manufacturing only after a customer order is in hand and a design is determined. Products offered can range from limited lines with few salable items to extremely large numbers of salable configurations. These factors all influence the architecture of data foundations and may require the use of planning or modular, generic, or activity data file organizations. These issues will be discussed in detail in Chapters 6 and 7.

For a company data file to be effective, the resulting information base must satisfy all users. Because everyone uses the company data foundations, those foundations must serve everyone's needs (see Figure 1-4).

The data foundations must also be accurate and complete. Completeness means that all required data are maintained. Accuracy means that the data being maintained are correct. The *garbage* out of the company system is often the result of the *garbage* that's in the company data files. Accuracy and completeness are discussed in Chapter 4.

Getting the data right is one requirement. Keeping the data right is another. For data foundations to support manufacturing functions adequately, good change control policies and procedures need to be an integral part of a company's operating philosophy. Everyone must realize the impact of wrong information and strive for accurate data maintenance and control. The same kind of control is required for a simple item change transaction, an engineering change update, or a new product introduction.

A solid manufacturing data foundation must

- represent the manufacturing process
- represent the products we sell and our strategies and plans for satisfying our customers

- be understood by and satisfy all users
- be complete and accurate
- be supported by good change control practice

Getting the data foundation right is not always an easy task. Change is difficult to implement because change to the data foundations typically affects all company functions. The scope of the change is often wide-

Figure 1-4
Functions of the Data Foundation

- General Manager
    - One Company Bill of Material
    - Data Accuracy
    - Effective Change Control
- Sales & Marketing
    - Sales Configuration
    - Product Options
- Purchasing
    - Supplier Schedule
    - Item Specification
- Engineering
    - Product Definition
    - Process Definition
    - Engineering Resource Planning
    - Design for Manufacturability
- Planning/MPS
    - Material Plan
    - Plant Schedule
- Manufacturing
    - Production Schedule
    - Shop Floor Control
    - Material Supply
    - Resource Plan
- Finance
    - Financial Planning
    - Manufacturing Cost and Accounting
- Service
    - Spares Plan
    - Configuration Control
- Information Systems
    - Software
    - Integration

spread. There is often considerable emotion and pride attached to the existing data structure. This emotion may be sparked when a simple, short numbering system is proposed that eliminates part number significance. Eliminating item numbers from drawings can create quite a stir. Agreeing on a single company data base approach when the *as designed* doesn't quite match the *as configured* can also generate a little excitement. Questions about what to include on bills of material and routings, whether or not to modularize, and even how to structure can cause lengthy debate that can be tough to resolve. Chapter 9 will discuss a proven process to implement change and overcome these emotional concerns.

Progressive manufacturers recognize that getting the data foundations right is a cornerstone of business success. Complete and correct data will provide a company with a single set of company information. Data debate is eliminated, and the focus can then be concentrated on problem solving, forward planning, and continuous improvement.

MAKING IT HAPPEN

By now, it should be clear how important a correctly structured data foundation is to supporting a company's objectives. Like a fine chowder of properly balanced ingredients, we must view the various functions of our companies as parts of a whole process. We must also realize, however, that the foundation of every great soup is a carefully prepared stock. Once that savory blended base is achieved, we should have few problems in satisfying the finicky appetites of those we serve.

What will follow, then, is a recipe for excellence. Our first step will be to define the manufacturing process, documenting all the necessary ingredients for a properly prepared data foundation. Once we have accomplished that, we will demonstrate how to structure bills of material and routings, making sure that the recipe is both accurate and complete, so that what is produced is exactly what the master chef has created.

With bills and routings now supporting the manufacturing process, we can then use them effectively to plan and control all required resources. We will also consider whether or not modularized bills of material are necessary to meet the planning and scheduling needs for the large number of options on the menu. We will also explain the process for introducing new products as well as the techniques necessary for handling engineering change control.

Once the basics have been presented, we will provide an understanding of some specific data foundation applications to help fine-tune the process to meet particular needs. Finally, we will demonstrate how to successfully implement these ideas into a coherent, proven data system designed to provide the required data foundation support to all functions in the company.

This is a working book, designed to be used, referred to, and referred back to again and again. The process begins, however, by being committed to becoming better. The information that will follow is dedicated to that objective.

Chapter 2

Defining the Manufacturing Process

"How does anything get done around here?" comes the plaintive cry of the vice president of operations as he addresses the managers of engineering, production, planning, purchasing, finance, and quality. "Everyone seems to be working at cross purposes. Can't we reach some kind of agreement so we can get some quality products completed when they're supposed to be, and can't we do it at a reasonable cost?"

"We'd be doing a better job of getting things out the door," the production manager says, defending his troops, "if engineering would only . . ."

"Oh, here it comes," contends the manager of design engineering. "If only engineering would what? We make sure our drawings are accurate, and any changes are sent down immediately. This whole place would fall apart if it weren't for our drawings. Why don't you pick on stores or planning for once. They're probably more to blame than us."

And the finger pointing continues for the next hour or so, with everyone ultimately leaving the meeting more upset than when they came in. Why can't they get their act together? From the sound of this meeting, there are probably a good many reasons. One reason, however, is that each functional area in the company appears to be working on its own individual requirements and priorities, relying on its own separate information.

Those of us in manufacturing must realize that before anything in a manufacturing environment can be built or shipped to consistently meet the needs of our customers, we have to establish a single data base that will support all of our varied information needs. The creation of this solid information foundation, as we mentioned in Chapter 1, becomes the cornerstone for all our operations. In this chapter, we will cover the basic data requirements necessary to adequately define a company's manufacturing processes.

ITEM OR PART NUMBERING

The first block in the construction of a manufacturing information base is the item number or part number, terms used interchangeably in this book.* These item identifiers define the material content of a company's product. Part or item numbers are assigned at the lowest levels in the manufacturing process, identifying purchased items and raw materials. They are commonly assigned at subsequent manufacturing levels such as blending, mixing, cutting, sewing, fabrication, or assembly. They may also be assigned at the highest level of manufacturing, including the identification of finished goods or end items. In addition, item identifiers can be used to define some nonmaterial requirements of our manufacturing processes, such as preventive maintenance activity, major tasks to support a product launch, product documentation, and tooling requirements. These will be discussed in later chapters.

Typically, a company will assign a part number to a material item whenever that item is purchased or stored. It will also assign one when the material is planned or scheduled with a work order as well as when it is necessary to link specific information to that material, such as a standard cost or an order policy.

It is not necessary, however, to assign part numbers for every step of the manufacturing process. In JIT/TQC environments, we often see parts identified as raw materials move directly through the entire manufacturing process with only the finished, salable item assigned a part number. Some progressive companies, using cellular manufacturing and other manufacturing flow techniques, have significantly reduced the

---

* Caution: Some companies maintain item numbers such that *part* #1234 can appear as *item* #15 on a bill of material and *item* #40 on the drawing. People familiar with this situation will not equate these two terms as do the authors of this book and the editors of the *American Production and Inventory Control Society Dictionary*. See the Glossary.

need for part numbers, identifying only a few subassembly or semi-finished levels of manufacturing. Other companies, however, because of traditional manufacturing methods, regulatory requirements, or customer requirements, have hundreds, thousands, and sometimes millions of item or part numbers on file.

A different part number needs to be assigned whenever one material is uniquely different from another and the two materials cannot be used interchangeably, such as sugar and salt. Generally, interchangeability is defined by company policy. One such policy states that an item is *interchangeable* when it (a) possesses functional and physical characteristics equivalent in performance, reliability, and maintainability to another item of similar or identical purposes and (b) is capable of being exchanged for the other item without alteration of the item or of adjoining items, except for adjustment or calibration.

Whenever a part change occurs in which the new part is not 100 percent interchangeable with the old part, the part number must be changed. An example of the need for unique identifiers can be found in a company that produces and sells felt pens. In this company, the specifications for ink colors are fairly simple. For blue, the specification calls for light and dark, and the ink is quality inspected against a color chart with a wide range of color tone allowed. Over the years, in fact, several color changes actually occurred that satisfied interchangeability rules with no impact on part number. But then the company had an opportunity to sell continuous, large quantities to one customer. However, this customer required a specific tint of blue, which was not to be sold to other customers. There were no other changes besides the ink tint specification. Immediately, a second part number was assigned because interchangeability was now a real problem.

To control unique item number assignment, companies need to establish rules of form, fit, and function:

1. *Form* defines the configuration of an item, including factors such as geometric shape, size, density, weight, or other parameters that uniquely characterize that item.

2. *Fit* is the ability of an item to physically interface or connect with or become an integral part of another item.

3. *Function* defines the results that an item is expected to perform.

Whenever the form, fit, or function of an item is changed, it is deemed no longer interchangeable, and a new part number is assigned. These definitions of form, fit, function, and interchangeability are offered as a guide. They might not be right for your company. It is important, however, that policies for item number assignment and their associated definitions exist, are clearly understood, and are supported in practice.

Many changes occur that don't affect the form, fit, or function of an item or its interchangeability. Common examples include a change in a specified tolerance, in a chemical specification, or in the alloy of steel used to produce an item. In these cases, a new revision level is usually assigned to document or note the new condition. Typically, revision levels are defined as an alphabetic character, with a new part assignment assigned an *A* level of revision. The revision level is not part of the part number and is only carried as an item descriptor on the item master file. In addition, each new revision is normally supported by an authorizing document, often a specification update.

One commonsense way to examine a company's control of unique item identification is the *blindfolded stockkeeper's test*. If we go to the stocking location for an item, blindly pick the material, and say, "Oops, it's red, not green," "Oops, that's an old version, and it's not interchangeable," or "Oops, that's powder, we need the capsule form," we're in trouble. Such a company's policies on unique part assignment need work.

Another condition that must be carefully controlled is stocking levels in the manufacturing process. Although our goal is to achieve flow in manufacturing, when we do stock items that are uniquely different from their prior manufacturing state, they must be assigned unique item numbers. This means that raw, semifinished, and finished materials, unformed and formed parts, and subassemblies and assemblies all need unique part numbers if they are to be stocked. A key question that every company should ask is, Do we *need* to stock all the intermediates we do today?

Another important point about item numbers is that they should be as short as possible. Part numbers are keyed, copied, and used as verbal identifiers. The shorter the numbers, the more accurate people can be. Obviously, the greater the number of digits in a part number, the greater the chance for error. We also recommend that only numeric digits be used. Alphanumeric schemes simply make the whole issue more prone to error. There's nothing like a few *l*'s or *I*'s to be confused with 1's. Or

O's to be confused with zeros. Or *Z*'s to be confused with 2's. A simple six-position numeric numbering system allows for one million parts. For our unfortunate friends with over a million parts on file, one more digit should not be a major problem.

Part numbers are often made longer by the use of significant digits. One company we know had a part number that looked like this: 123 A 45678. In this configuration, the first three numeric digits categorized the type of material. The letter signified whether it was manufactured, purchased, or a piece of hardware. The last five numbers were nonsignificant identifiers. The first problem with this format was accuracy. Anyone who has worked with a computer keyboard knows how easy it is to make a mistake going from the numeric keypad to the alpha keyboard and back. The second problem was significance. After a year or two, the engineers in this company became confused on the definition of *type of material* and began to use rubber and plastic instead of metal for making brackets, pulleys, and panels. Before long, the first three characters became nonsignificant because the definition was no longer clear. Furthermore, no one cared. The numbers that some people thought were significant at one time were no longer significant. Like most part-numbering schemes, this one didn't last.

The same company also went through a major change in the way it did business. In a move to reduce cost, they found that they could purchase parts less expensively than they could manufacture them. Initially, they tried to maintain the letter significance in the item number and processed a part number change with every make/buy revision. It didn't take long to recognize this as an unnecessary, costly effort, and the alpha character became nonsignificant. Because of the cost to rewrite all the programs that used the part number, this company still maintains the nine-position alphanumeric part number, even though today it is totally nonsignificant.

Prior to computerization, significant part numbers made sense. They allowed us to tell things about the material used in our products, such as material type, commodity code, or make/buy codes. But as we have seen, what was once significant often loses its meaning over time. As conditions change, significant identifier schemes break down. More important, using computer fields, we can now maintain all kinds of desired information on the item master file identified with a simple part number. Such significant data can be easily accessed and reported.

A word of caution is required. In many businesses, the item number used to identify a specific customer configuration can be very significant. For example, in the office furniture business, customers can configure a steel office desk from a sales catalog by creating an item number as they select numbered options such as type of material, number of drawers, types of hardware, and size. This area will be discussed further in Chapter 6.

Item number significance can work well at the component level. Companies will sometimes suffix a base part number with a "dash number" as part of their part-numbering scheme. An aluminum window company used approximately 100 different cross sections of aluminum extrusions. Each had a unique item identifier. As each of these extrusions was cut to different lengths, it took on a unique form. Uniqueness was maintained by adding a two position number to the base extrusion part number, giving the actual material length. Similar applications of dash numbers have been used to describe color variations and in remanufacturing to distinguish between a returned "core," refurbished, or new part. Such simple, visual use of item number significance works well. More complicated uses tend to fail over time and cause data maintenance errors.

Changing part number formats may not be an easy process. To accomplish this, most companies try to develop a phased approach, or they begin with new products. An electronics company phased in a new part number format down through the bill of material levels beginning with its product catalog. A major producer of food containers with several plants phased in a new format as each site installed new manufacturing systems.

Many companies in transition will also create cross-reference lists of old and new item identifiers. This approach needs to be used carefully, however, as the cross-reference process can become a crutch that everyone depends on and that is never eliminated. This usually results in some real maintenance problems and is often the source of errors.

Making certain that you have properly defined your item-numbering system is the initial step in the data foundation process. Once that is in place, the next foundation layer can be laid, which contains the data needed to support that item number.

ITEM MASTER DATA FILE

Information that was once contained in significant item numbers can now be found in the item master data file. The item master file describes the item number by listing all the information that is true about that particular item in all its applications. Quantity per is not found on item records as different amounts of the item may be needed in one assembly versus another.

The item master file might contain the following kinds of data:

- Part description
- Engineering drawing number
- Revision level
- Commodity code
- Cost(s)
- Product code
- Lead time
- Order quantity
- Item type or status
- Latest engineering change dates
- Lot size
- Planner/buyer codes
- Unit of measure
- Make or buy

DRAWINGS OR SPECIFICATIONS

Companies that produce fabricated and assembled products use drawings to effectively communicate a part's pictorial description and to convey its dimensions and tolerances. Prior to computerization, parts lists were also maintained on the drawings—a practice still continued by some companies today.

However, with computerization, most companies have discontinued this practice. The bill of material information maintained in the computer data base is used to eliminate the list of part numbers on drawings. More important, the waste of dual maintenance and the problems of keeping two different sources of information have also been eliminated. Only when regulations or customer needs call for components on drawings should such a practice be supported. And even in these situations, a copy of the bill of material accompanying the drawing will often satisfy the customer or regulatory need.

The technique of *ballooning* items on a drawing provides a linkage between the computerized bill of material and the drawing. The *balloon* or *find* number from the drawing is carried as a data element on the product structure or bill of material record. A bill of material that ties a balloon number to each component makes it easy to identify the item numbers specified on a drawing.

Manufacturers of other types of products, including many process industries, may maintain specifications, formulations, or other technical information rather than engineering drawings. It should be noted that the same potential to maintain duplicate material requirements information can exist in these environments as well. It is important to work toward a single source of information.

Technology is having a great impact in the drawing office. Today, drafting is an integral part of the engineering design process, with drawings created at computer-aided design (CAD) workstations. In some companies, group technology is eliminating some design effort altogether, using previously designed parts to satisfy new material requirements.

The issue here is that, once again, with parts lists on CAD, the components are coupled with the drawing. These components need to be investigated for manufacturing process, possibly restructured, or duplicated onto the company data files. Many companies are streamlining this process, and some are even electronically downloading data from specialized CAD systems to company files. This will be discussed further in Chapter 7.

BILLS OF MATERIAL

The bill of material describes the material content of a product at each stocking level in the manufacturing process. The item being produced is called a *parent*, and the materials required to produce the item are *components*. The components are listed with their required usage or *quantity per* necessary to make one parent item. When these one-level, or *single-level,* parent/component relationships are established, the computer can plan, cost, and report all types of single and multilevel

Defining the Manufacturing Process 25

information. In some companies, bills of material are actually called *recipes*. In others, they are identified as *formulations* or *specifications*. Bills of material may also include nonmaterial component items, such as instructions or consumable tooling required to support a parent's manufacturing process.

To better understand these parent/component relationships, consider the CCS Carpet Cleaning Company and its number-one-selling item— item #0123, the carpet cleaner shown in Figure 2-1.

**Figure 2-1
Carpet Cleaner**

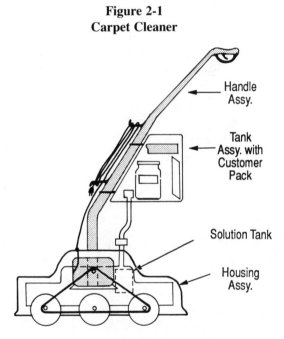

The #0123 cleaner is made up of the components identified in Figure 2-2. Now, it may be obvious that it takes a few more parts than these to make a carpet cleaner. For our purposes here, however, this list represents everything that goes into the product.

What this engineering parts list does not tell us is what the parent/component relationships are in the manufacturing of this product. To understand the manufacturing of a #0123, we need to see the single-level relationships.

Figure 2-2
Engineering Parts List

Parent: 0123
Description: Carpet Cleaner

COMPONENT	DESCRIPTION	UNITS REQUIRED	UNIT OF MEASURE
0403	Clamp	1	Ea.
1115	Customer Pack	1	Ea.
1196	Housing	1	Ea.
1201	Gasket	2	Ea.
1910	3-Gal. Tank Subassembly	1	Ea.
1959	Instruction Set	1	Ea.
2156	Trigger Assembly	1	Ea.
2927	3-Gal. Tank Assembly	1	Ea.
3215	Solution Tank	1	Ea.
3219	Handle	1	Ea.
3804	Housing Assembly	1	Ea.
4209	Painted Tank Top	1	Ea.
4315	Brush Assembly	3	Ea.
4651	Handle Assembly	1	Ea.
5319	Valve Assembly	1	Ea.
5640	Steel	10	Lbs.
5704	3-Gal. Painted Tank Bottom	1	Ea.
5746	Hose	10	In.
6111	1-Hp. Motor	1	Ea.
6221	Bottled Concentrate	1	Ea.
7114	Power Cord	1	Ea.

The components needed to construct a standard #0123 carpet cleaner are a handle assembly, tank assembly, housing assembly, and customer pack (see Figure 2-3). Using this single-level bill of material structure, we see the actual material content of the #0123 carpet cleaner. These data are critical for material planning, product costing, and in defining how to construct the product. This structure allows the computer to plan matched sets of components to support production schedules, determine lower-level material requirements, issue pick lists, and roll up material cost.

Multilevel bills of material are then composed of a series of single-level structures. Since the #0123 carpet cleaner requires a #2927 tank assembly, it is a simple matter to retrieve the single-level bill of material for the #2927. If the single-level bills of material are correct in the computer, multilevel structures are easily determined and also correct (see Figure 2-4).

Figure 2-3
Single-level Bill of Material

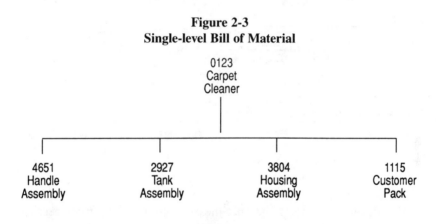

Single-level
Bill of Material
Retrieval

Parent: 0123
Description: Carpet Cleaner

COMPONENT	DESCRIPTION	QUANTITY PER
1115	Customer Pack	1
2927	Tank Assembly	1
3804	Housing Assembly	1
4651	Handle Assembly	1

Figure 2-4
0123 Carpet-Cleaner Levels

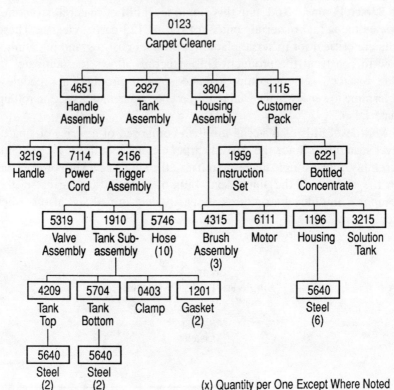

(x) Quantity per One Except Where Noted

Multilevel structures are sometimes known as *family trees* because they are pictured similarly. If we call the end item cleaner level 0, components used in the final assembly level would be level 1, and their immediate components would be level 2. The components of the #1910 tank subassembly would then be level 2; the #5640 steel going into the housing would be level 3; the #5640 steel required for the manufacture of the tank top and bottom would be level 4.

Although it is possible for computer software to draw these family trees, the standard way of displaying multilevel bill of material structures is the indented parts list report or screen (see Figure 2-5). Data are retrieved and organized in such a manner that our single-level and

Defining the Manufacturing Process 29

multilevel relationships are quite clear. The four components that go into the #0123 carpet cleaner are easily determined by the level 1 identifiers. It is also easy to see that the #3804 housing assembly takes four components at level 2, and that one of those level 2 components, #1196 housing, has a level 3 component—#5640 steel.

CCS also manufactures and sells a cleaning fluid concentrate. This process type of manufacturing is considerably different from the fabrication and assembly operations of the carpet cleaner plant, but the supporting systems are very similar. The data foundations use exactly the same data files, and both areas use the same software. The "family tree" of the carpet cleaning concentrate is illustrated in Figure 2-6. The

Figure 2-5
Indented Bill

Parent: 0123
Description: Carpet Cleaner

LEVEL	COMPONENT	DESCRIPTION	UNITS REQUIRED	UNIT OF MEASURE
1	1115	Customer Pack	1	Ea.
• 2	1959	Instruction Set	1	Ea.
• 2	6221	Bottled Concentrate	1	Ea.
1	2927	3-Gal. Tank Assembly	1	Ea.
• 2	1910	3-Gal. Tank Subassembly	1	Ea.
• • 3	0403	Clamp	1	Ea.
• • 3	1201	Gasket	2	Ea.
• • 3	4209	Painted Tank Top	1	Ea.
• • • 4	5640	Steel	2	Lbs.
• • 3	5704	3-Gal. Painted Tank Bottom	1	Ea.
• • • 4	5640	Steel	2	Lbs.
• 2	5319	Valve Assembly	1	Ea.
• 2	5746	Hose	10	In.
1	3804	Housing Assembly	1	Ea.
• 2	1196	Housing	1	Ea.
• • 3	5640	Steel	6	Lbs.
• 2	3215	Solution Tank	1	Ea.
• 2	4315	Brush Assembly	3	Ea.
• 2	6111	1-Hp. Motor	1	Ea.
1	4651	Handle Assembly	1	Ea.
• 2	2156	Trigger Assembly	1	Ea.
• 2	3219	Handle	1	Ea.
• 2	7114	Power Cord	1	Ea.

30   MANUFACTURING DATA STRUCTURES

indented bill of material report shown in Figure 2-7 is similar to the report used in the carpet cleaner plant.

To achieve the right amount of component usage, decimal quantities are used. This allows component planning to be correct regardless of the parent requirement that is scheduled and also allows the company's costing systems to roll up to the correct product cost summary. At CCS, the software allows for enough decimal places to adequately serve their needs.

WHERE-USED REPORTING

Sometimes it is important to know not only what components are required to produce a parent but also where a component is used in all of its higher relationships. Once single-level parent/components are correctly identified in the company's data files, such *where-used* information is easily retrieved. A single-level, where-used report for CCS item #4209—painted tank top—shows that it is required in the production of three next higher subassemblies (see Figure 2-8).

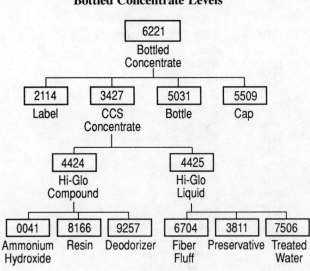

Figure 2-6
Bottled Concentrate Levels

Figure 2-7
Indented Bill of Material

Parent: 6221
Description: Bottled Concentrate

LEVEL	COMPONENT	DESCRIPTION	UNITS REQUIRED	UNIT OF MEASURE
1	2114	Label	1.00	Ea.
1	3427	CCS Concentrate	1.00	Litre
• 2	4424	Hi-Glo Compound	50.00	Grams
• • 3	0041	Ammonium Hydroxide	15.00	Grams
• • 3	8166	Resin	30.00	Grams
• • 3	9257	Deodorizer	5.00	Grams
• 2	4425	Hi-Glo Liquid	1.00	Litre
• • 3	3811	Preservative	0.02	Litre
• • 3	6704	Fiber Fluff	0.02	Litre
• • 3	7506	Treated Water	0.80	Litre
1	5031	Bottle	1.00	Ea.
1	5509	Cap	1.00	Ea.

Figure 2-8
Single Level: Where-Used

Item: 4209
Description: Painted Tank Top

LEVEL	PARENT	DESCRIPTION	QUANTITY PER	UNIT OF MEASURE
1	1910	3-Gal. Tank Subassembly	1	Ea.
1	4721	4-Gal. Tank Subassembly	1	Ea.
1	7350	5-Gal. Tank Subassembly	1	Ea.

Often, it is useful to see both the next higher use of an item and all higher uses of that item, including end items. A multilevel where-used look at #4209—painted tank top—shows that it is always used at level 3 and is used in three carpet cleaner products (see Figure 2-9). Both single-level and multilevel where-used retrievals are very useful in CCS's concentrate division, since several of their chemicals have many next higher applications.

Figure 2-9
Multi-level: Where-Used

Item: 4209
Description: Painted Tank Top

LEVEL	PARENT	DESCRIPTION	QUANTITY PER	UNIT OF MEASURE
2	1910	3-Gal. Tank Subassembly	1	Ea.
• 1	2927	3-Gal. Tank Assembly	1	Ea.
• • 0	0123	Carpet Cleaner	1	Ea.
2	4721	4-Gal. Tank Subassembly	1	Ea.
• 1	1527	4-Gal. Tank Assembly	1	Ea.
• • 0	0124	Carpet Cleaner	1	Ea.
2	7350	5-Gal. Tank Subassembly	1	Ea.
• 1	2104	5-Gal. Tank Assembly	1	Ea.
• • 0	0125	Carpet Cleaner	1	Ea.

OTHER BILL OF MATERIAL USES

In addition to defining specific parent/component relationships, there are other uses of bills of material. In custom product companies, where every product is designed from customer requirements, and in new product introduction planning, a *generic bill of material* is often used. This is an organization of items that represents how a product is generally structured. It's used primarily for forward planning when specific requirements aren't known. These bills are often tied to *generic routings* for forward capacity planning.

Some companies also use *bills of activity*. These bills describe a series of dependent steps or actions similar to a network project plan. They're used by companies to plan areas like design engineering, tool acquisition, and documentation. They are often combined with material structure to give a complete planning capability for situations like a new product launch. As with generic bills, bills of activity are usually tied to routings to plan capacity requirements associated with the structured activity. Generic bills and bills of activity will be discussed further in Chapter 7.

In companies involved in remanufacturing operations, *ratio* or *percentage* bills of material are utilized to plan the estimated use of new and refurbished materials in the rebuild operations. These types of bills are discussed in Appendix C.

A fourth type of bill is the *planning* or *super* bill of material. This bill organizes product features and options within a planning family in a manner using percentage quantity per relationships to forecast anticipated use. Planning bills will be described in greater detail in Chapter 6.

Phantoms and Pseudos

Two forms of bill of material coding and structuring require explanation at this point. The first is a *phantom*, or *transient*, bill of material item. *Phantoms* are items produced in the manufacturing process and thus are definable parent items, but they are not typically stocked. Instead, they physically remain in the manufacturing process and are quickly consumed as components in their next higher level item.

An example of a phantom is the engine assembly and the transaxle assembly that, as components, become an engine/transaxle assembly during automobile assembly. The engine/transaxle assembly is not built in batch or stocked. Rather, it is built on a feeder line in a flow and is almost immediately assembled into an automobile. Blended dough pulp ready to be placed in the oven is an example of a phantom level in a recipe.

The value of the phantom is in material planning, which is discussed in Chapter 5. Although specifics differ in various software, material requirements planning (MRP) usually recognizes a phantom and plans its requirements, but does not create a phantom plan or schedule. Instead, the system "blows through" the phantom level and plans the components in support of the phantom's next higher level requirements. In our automobile example, engines and axles would be planned and scheduled for automobile production. The planning system simply "blows through" the phantom engine/transaxle assembly level. If a phantom happened to be in stock, perhaps because of a field return or an imbalanced quantity of production, the planning system would recognize its availability.

Phantom items are generally assigned zero lead times and lot-for-lot order quantity and may also be coded with a phantom item type. As companies improve their factory flow and eliminate intermediate stocking levels, the use of phantom codes eliminates the need to restructure bills of material. Rather than eliminating the parent that is no longer required, and restructuring all the components into the next higher assembly, that parent is simply coded as a phantom.

Pseudos are artificial groupings of components useful for planning.

34   Manufacturing Data Structures

Unlike phantoms, however, the components of a pseudo cannot be manufactured together into a stockable parent configuration. For this reason, pseudos can never be inventoried. Pseudos are treated like phantoms by material requirements planning. Again, the planning system will blow through the pseudo and plan its components in support of next higher demand.

An example of a pseudo is the group of all the common parts required to assemble option/feature-driven assembled products. Pseudos and phantoms are discussed further in Chapter 6.

Routings and the Work Center Master File

So far in our definition of the manufacturing process, we have been discussing material content. The bill of material describes the level-by-level structuring of that material to produce a product. What has not been discussed is how the product gets produced. This is the role of the routing.

The *routing* defines the specific step-by-step method of manufacture necessary to take a component, or set of components, and produce a parent. For every parent item in the manufacturing process, a bill of material and routing are maintained. The routing detail is arranged in the order or sequence of the single-level manufacturing process. Each step is assigned a number or *operation* identifier. For each operation, the department and work center where the work is to be performed are identified; the expected hours of machine setup, equipment changeover, or cleanup time is described; and the expected time to complete the operation for a single item (or piece), known as the operation *standard*, is described. Additionally, some routing files allow users to maintain other support information, such as the labor grade required to perform the work and any tooling or fixtures required to support the operation process. Figure 2-10 shows the routing for CCS's tank bottom, #5704.

Supporting and linked to the routing file is a work center master file. This file contains data about each work center, such as the time it takes to move material to or from that work center, the amount of work, or *queue*, that is planned in order to maintain efficient operations, the historical or demonstrated output or capacity of the work center, and the labor and/or equipment factors used in planning capacity. Alternate work centers may need to be defined, as will alternate routing steps, in some companies.

Figure 2-10
Routing

Parent: 5704
Description: Tank Bottom

OP. NO.	DEPT.	W.C.	DESC.	SU.	RUN	LABOR	TOOL
10	FAB	120	Cut Tank Blank	2.0	.010	16	1824
20	MACH	240	Drill Hose Inlet	1.0	.020	13	—
30	FAB	160	Form Gasket Ring	2.0	.010	16	1893
40	FAB	160	Form Box	2.0	.010	16	1762
50	WELD	370	Weld Box	0	.030	18	—
60	FAB	110	Deburr as Necessary	0	.005	04	—
70	INSP	490	Inspection-Instruction 10460	2.0	0	11	—
80	PNT	560	Paint	0	.020	10	—
90	INSP	410	Inspection-Instruction 10390	0	.002	14	303

Routings provide the required manufacturing data on how a product is produced in all types of manufacturing operations. At CCS, routings are found in both the concentrate process plant and the carpet cleaner fabrication and assembly operation. The data maintained in the routing file and in the work center master file are used to plan capacity, schedule factory operations, and cost work-in-process, inventory, and factory operations.

OTHER SUPPORTING DOCUMENTATION

The item master file, bill of material, and routing for an item contain a lot of information describing that item, its material content, and the process used to make it. However, additional information not normally maintained in these files is always needed. This information might include physical properties, such as hardness of a wearing part that is heat treated, operating temperatures, or the pressures for a textile dyeing process.

Most manufacturers maintain this necessary information in some auxiliary way. Master documents such as *manufacturing instructions* may be maintained and reproduced as hard copy when needed. Word processing techniques are often used to facilitate maintenance. Some companies utilize *text* or *comment* files provided by manufacturing software systems.

Regardless of the approach used, it is important to recognize that supporting information must be organized so that it easily integrates with the data base. Data contained on the item master, bill of material, routing, and work center files are considered the base or root definition; other information is supplemental.

Many companies, notably in process industries, combine data base information from bills of material and routings as well as supplemental information and produce a consolidated document or retrieval called a variety of names—*manufacturing process instruction*, *batch card*, *recipe card*, *construction sheet*, *mix ticket*, or *melt order*. It's usually not evident where the actual information was originally stored.

THE *LIVE* DATA FOUNDATIONS

We have discussed the item master file, which describes the fairly constant, or static, information about an item number; the bill of material file, which describes the planned material content at the various levels of the manufacturing process; and the routing file, which describes the planned processes used to produce the products to be manufactured.

These files define and describe a company's items, products, parent/component relationships, and manufacturing processes. For the most part, the information on these files is reasonably stable—or static—and file maintenance is carefully controlled. For this reason, these files are often referred to as the *master files*. They are the main reference files of information.

A manufacturing company needs master files to plan and schedule its operations and to plan its future. In executing day-to-day operations, however, additional information is required that represents the actual status of materials and processes in the plant. For this reason, another set of data foundation files, known as *live files*, exists to track and maintain this more dynamic information.

The first of these files is the *inventory file*, or inventory record, where the on-hand and on-order balance of materials is maintained. As in a personal checkbook, every inventory and order transaction posts to this file, allowing the requirements planning system to recognize supply against demand and allowing the financial systems to account for inventory value.

The second live file is the *requirements file*, or *allocation file*. As each production order or schedule is established, its material content is determined and posted to a requirements file. Although the component requirements will usually be the same as those specified on the *master* bill of material, actual production may require the use of a substitute item. When the order is actually released, the requirements file is utilized to document all actual material usage, to report the current material status of an order, to collect actual or standard material cost to an order, to recognize actual versus planned material variance, and to record certain traceability data for historical retention. The use of the requirements file is further discussed in Chapter 5.

The third live file is the *operations detail file*, or *work-in-process detail file*. Like the requirements file, the operations detail file is a record of all established production orders. The operations detail file tracks actual routing flow, including the use of any alternate routing operations. The file is used to report the current operation-by-operation status of an order, to collect actual or standard production cost to an order, to recognize actual versus planned production variance and labor efficiency, and to record certain traceability data for historical retention. The use of the operations detail file is also discussed in Chapter 5.

THE MANUFACTURING ENVIRONMENT: JOB SHOP VERSUS FLOW SHOP

Certain characteristics of a company's manufacturing data foundation (bill of material, routings, etc.) relate to the specifics of the company's manufacturing environment. One critical aspect of the environment is the degree of *job shop* versus *flow shop*.

Here are some current definitions of the American Production and Inventory Control Society (APICS). While these definitions are regarded as the most widely accepted standard, note the substantial overlap:

- *Intermittent production*—A form of manufacturing organization in which the productive resources are organized according to function. The jobs pass through the functional departments in lots, and each lot may have a different routing. Syn.: Job shop.

38   Manufacturing Data Structures

- *Continuous production*—A production system in which the productive equipment is organized and sequenced according to the steps involved to produce the product. Denotes that material flow is continuous during the production process. The routing of the jobs is fixed, and setups are seldom changed. Syn.: mass production.

- *Flow shop*—A form of manufacturing organization in which machines and operators handle a standard, usually uninterrupted, material flow. The operators generally perform the same operations for each production run. A flow shop is often referred to as a mass production shop, or is said to have a continuous manufacturing layout. The plant layout (arrangement of machines, benches, assembly lines, etc.) is designed to facilitate a product "flow." Some process industries (chemicals, oil, paint, etc.) are extreme examples of flow shops. Each product, though variable in material specifications, uses the same flow pattern through the shop. Production is set at a given rate, and the products are generally manufactured in bulk.

- *Process manufacturing*—Production that adds value by mixing, separating, forming, and/or chemical reactions. It may be done in either batch or continuous mode.

- *Repetitive manufacturing*—A form of manufacturing where various items with similar routings are made across the same process whenever production occurs. Products may be made in separate batches or continuously. Production in a repetitive environment is not a function of speed or volume.

Plant layout, organization, and job shop versus flow shop characteristics have significant implications for manufacturing data foundations. Bills of material tend to be *deeper*—have more levels—in the job shop environment. As noted in Chapter 1, there is frequently more lot sizing for efficiency of manufacture, which results in intermediate stocking, item numbers, and hence more bill of material levels. Similarly, the functional layout of the job shop leads to *deeper* (longer) routings, with more detailed operation steps planned for individually scheduled work centers.

The layout and design in a flow shop usually serve to simplify both the bills of material and the routings. Even a mass production facility producing complex assembled products with hundreds and thousands of components, as in automotive assembly, usually operates with surprisingly *shallow* bills of material. Although several levels of subassembly may actually be produced, they are treated as phantom or transient levels of manufacture and only exist for moments before being consumed in the overall assembly process. Very detailed and precise manufacturing process information may be, and usually is, used to design the manufacturing flow, but minimal routing detail is needed for actually planning and scheduling factory operations. The sequence and flow are predetermined by the design of the manufacturing process. As an exaggerated example, an entire assembly line can be viewed as a single work center for planning and scheduling purposes.

The introduction of cells into the job shop environment or the development of continuous batch processes in a process environment directly leads to simplified routings. In these environments, equipment used frequently to perform sequences of individual operations is grouped into a manufacturing cell. The work performed in the cell is sometimes identified as a single operation—"make complete."

Two companies, one organized as a job shop and the other as a flow shop, will have dramatically different manufacturing lead times to produce the same item. Much, if not all, of that difference is due to the *queue*, or wait time, incurred in the job shop, because work does not flow and must await access to the next work center.

The routing capabilities found in most standard MRP II software may not truly permit effective simulation of a flow process in detail. For example, substantial overlap of detailed operations actually occurs in most flow processes. Conventional routing design often assumes that each detailed operation is performed completely before work moves to the next operation. Provisions made for operation overlap may be too crude to truly reflect reality, and custom logic may be needed to solve the problem.

This weakness may not be a problem in terms of effectively planning and controlling the flow. The detailed execution tools of MRP II, input/output control, and the dispatch lists of shop floor control, which are virtual musts in the job shop, may not be needed at all in the flow shop. Similarly, detailed capacity requirements planning may not be needed. A well-designed, although admittedly normally more extensive, rough-cut capacity planning process is usually adequate.

Finite scheduling software is more prevalent in flow shop environments than in job shop environments. One reason for that is probably the difficulty in modeling the flow shop with routing and work center techniques designed for the job shop. Another reason is the usually simpler scheduling problem of the flow shop versus the job shop.

MULTIPLANT CONSIDERATIONS

Multiplant manufacturers as well as distribution networks that use a single data base for their businesses must consider location coding information. In these environments, the same item might be stored in or distributed to several locations, or one plant may supply an item to other locations. An item might also be produced in several locations with different bills of material and/or routings.

It is usually necessary to segregate information by location, such as on-hand, on-order, and in-transit inventories, costs, lead times, make-buy codes, and order policies. One approach commonly used to identify an item in a given location, an *item/location*, is to add a prefix or suffix (a location code) to the item or part number. Some companies simply reserve one or two positions of the item or part number field for the location code. When one is provided, a separate field for the location is more flexible. Relational data bases make this less of an issue today.

Some multiplant software handles location coding without intervention from the user. Item numbers are automatically updated with location codes for processing and reporting and are not seen by the user. Other software requires location coding to be externally entered. Comprehensive software will also provide detailed information by item/location as well as various consolidated reporting and retrievals.

Planning linkages between receiving and supplying locations are created in different ways. Bills of material are created to link the receiving location to the supplying location. Item A in location 1 (receiving location) is the parent of component item A in location 2 (supplying location). Processes such as material requirements planning use these bill of material linkages to place demands on the supplying location.

A second method used to represent the product supply network is to maintain a source code on each item/location. Separate computer programs are used to project demands from each receiving location on the supplying source.

Single data base designs for multiplant applications easily become complex and can quickly create deep, multilevel bills of material. The economics of distributed computing and advances in computer network-

Defining the Manufacturing Process   41

ing software can provide attractive alternatives to single data base designs.

Taking software designed for single-location use and adapting it to multiplant use requires work. Consolidated reporting/retrieval needs to be developed. In-transit material tracking is normally necessary and usually requires enhancements to the schedule receipts system to track needed shipment data. Data security considerations can be significant.

"AS-DESIGNED"/"AS-MANUFACTURED" CODING

In many companies, the design or engineering view of a bill of material is different from the manufacturing or planning view. Often, manufacturing needs to add intermediate or subassembly levels in support of the manufacturing process, resulting in levels in the manufacturing bills that are not part of the original design. Other differences occur as the design evolves over time.

At CCS this problem was addressed through component coding on the bill of material file. Here are the actual bill of material codes that were used:

CODE	USE
ACT	Current active configuration
CAN	Previous configuration, now canceled
PLN	Future configuration
SPC	Previous configuration where component is active for special use, such as spares
ENG	Engineering configuration only
MFG	Manufacturing configuration only

In Figure 2-11, the CCS bill of material for the #2927 tank assembly is displayed with its departmental coding. An individual reading this bill of material, or the computer system, would determine active *as-designed* or *as-manufactured* structure by reading the active codes in the engineering and manufacturing fields on the file (see Figure 2-12).

Figure 2-11
Departmental Bill of Material Coding

Parent: 2927
Description: Tank Assembly

COMPONENT	DESCRIPTION	Q./P.	DEPARTMENTAL CODING ENG.	MFG.
0403	Clamp	1	Act.	Eng.
1201	Gasket	2	Act.	Eng.
1214	Gasket	1	Pln.	Eng.
1910	Tank Subassembly	1	Mfg.	Act.
4209	Tank Top	1	Act.	Eng.
5319	Valve Assembly	1	Act.	Act.
5704	Tank Bottom	1	Act.	Eng.
5720	Hose	1	Can.	Spc.
5746	Hose	1	Can.	Act.
5790	Hose	1	Act.	Pln.

The two configurations are different because manufacturing has added a subassembly—the #1910 tank subassembly—as part of the build process. Engineering views the design as a single-level bill structure, while manufacturing views and actually assembles the tank assembly in a two-level bill of material process.

Examining the departmental coding, some other information can be determined:

- A #1214 gasket is in the design phase.

- A #5720 hose is no longer in production, but remains active for special use.

- A #5790 hose is being phased in, replacing the #5746 hose now active in manufacturing.

At CCS, the departmental coding maintained by the data administration group is based on engineering change documentation distributed by the engineering change coordinator. The data processing group is currently working to attach effectivity dates to the codes to identify configuration timing as well as configuration status. CCS has also created a

Defining the Manufacturing Process 43

Figure 2-12
Current Active Structure

"As-Designed"

Parent: 2927
Description: Tank Assembly

COMPONENT	DESCRIPTION	Q./P.	ENG.	MFG.
0403	Clamp	1	Act.	Eng.
1201	Gasket	2	Act.	Eng.
4209	Tank Top	1	Act.	Eng.
5319	Valve Assembly	1	Act.	Act.
5704	Tank Bottom	1	Act.	Eng.
5790	Hose	1	Act.	Pln.

"As-Manufactured"

Parent: 2927
Description: Tank Assembly

COMPONENT	DESCRIPTION	Q./P.	ENG.	MFG.
1910	Tank Subassembly	1	Mfg.	Act.
5319	Valve Assembly	1	Act.	Act.
5746	Hose	1	Can.	Act.

. . . AND . . .

Parent: 1910
Description: Tank Subassembly

COMPONENT	DESCRIPTION	Q./P.	ENG.	MFG.
0403	Clamp	1	Mfg.	Act.
1201	Gasket	2	Mfg.	Act.
4209	Tank Top	1	Mfg.	Act.
5704	Tank Bottom	1	Mfg.	Act.

history file in which fully canceled components are retained for historical tracking.

In one company we know, which manufactures power tools, a third configuration—*as-customer-documented*—is maintained by a similar bill of material coding technique.

Although this approach would work for many companies, it probably would not apply to situations such as those found in aircraft manufacturing, where complete sets of drawings and components to assemble total avionics, hydraulic, and airframe systems are issued. The manufacturing process normally assembles portions of the aircraft in stages, such as a wing assembly, that combines several elements of the as-designed documentation—avionics, hydraulics, and airframe. Component coding to develop a combined company data base might be a waste of time, and in such situations, there is usually a direct comparison completed at which time any as-designed/as-manufactured variance is reconciled.

One company in the smaller commercial aircraft market compared both configurations in their data base and completed a mechanical part-by-part match. They then documented the authorization for any mismatch condition. Copies of the comparison reports were provided as part of the customer documentation.

SOFTWARE CONSIDERATIONS

Generally, the software available today adequately supports the basic data base requirements of defining item or part numbers, material content, and manufacturing process. Maintenance capabilities typically include *same-as-except* features with add/delete maintenance screens and single-item or mass update facilities. On-line editing is common. Many packages allow some form of alternate bill of material or routing records on the master files.

With properly structured data foundations, almost any type of reporting is possible. This includes standard reporting, such as single-level and indented bills or where-used reports and screens. Specialized and customized reporting, such as batch specifications that might include item master, material content, and manufacturing detail information, are usually user-defined and -developed capabilities.

The problem with today's software packages is in their use, or rather lack of use, of the basic data files in their application programs. Needed capabilities, such as backflushing, forward lead time offset, customer order configuration, special bill coding, and other requirements, all discussed later in this book, are sometimes not available in the packages on the market. As these requirements are discussed, we will identify the software features required. *MRP II Standard System,* available from Oliver Wight Limited Publications, Inc., is also an excellent guide to defining your software needs.

MAKING IT HAPPEN

The data foundations of a manufacturing company are as basic as the structural foundations of the building in which the manufacturing takes place. No one would construct a building with weak materials and risk collapse. Yet, in many companies, basic manufacturing information is missing, incomplete, or inaccurate. Solid data foundations mean that our materials are identified, the material content of our products is properly structured, and our specific manufacturing process steps are described with a high degree of accuracy. The data base must support the organization's requirements so that the business can really run on one set of numbers.

Defining our manufacturing processes correctly is the key to successful planning, scheduling, and business control systems. Manufacturing Resource Planning fails every time when the data are incomplete or inaccurate. JIT/TQC also requires solid manufacturing information to foster improvement opportunity.

The first step in making it happen requires an understanding of the basic data foundation requirements described in this chapter. With that understanding, we are prepared to address the more specific issues of structure, accuracy and completeness, planning/scheduling support, modularization and change management, which are discussed in the following chapters. Chapter 3 specifically addresses the issue of properly structuring a company's manufacturing process information.

Chapter 3

Defining Levels in the Bills of Material and Routings

Whether we're preparing bouillabaisse in a restaurant or deciding how to construct the bills of material for a company that makes carpet cleaners, we need to define the various levels of our products and processes. This will basically be determined by how we treat semifinished or intermediate items as manufacturing progresses from raw materials and purchased components to finished product. A *level* is created in a bill of material when a parent/component relationship is defined. In a similar way, a routing *step* is created when an operation/work center relationship is defined for a parent item or part.

If we look at the process of making bouillabaisse, our list of ingredients, or parts list, would total 20 items and look something like the list in Figure 3-1. We might decide, however, that to maintain a strict adherence to our process, we want to make sure that each process step is properly reported. Our chef would go to the refrigerator and issue himself an onion, report that, chop it, then return it to the refrigerator in the location for chopped onions with its accompanying new part number, and report the transaction with a stock receipt. He would then follow this same procedure with the leek and fennel. Once that was completed, he would issue some oil, pour it into a Dutch oven, return the bottle to the cupboard, and issue out the chopped onions, leek, and fennel, and add them to the oil. While that mixture was sautéing, he could return to the refrigerator, issue the garlic, mince that, return it to the refrigerator, retrieve an orange, peel it, return it to the refrigerator, and report all transactions. He could then issue himself the minced

Figure 3-1
Bouillabaisse

This highly seasoned fish stew comes from the Mediterranean coast.

Prep. time: 30 minutes
Cooking time: 50 to 60 minutes

---

3 tablespoons extra-virgin olive oil
1 cup chopped onions
½ cup chopped leek
1 bulb fennel, chopped
1 tablespoon minced garlic
1 strip (3 in.) orange peel
¼ teaspoon thyme
Pinch saffron threads
2 cans (14 oz. each) tomatoes, drained
4 bottles (8 oz. each) clam juice

1 cup water
½ cup dry white wine
½ teaspoon salt
1 dozen littleneck clams, scrubbed
12 ounces monkfish fillets, cubed
12 ounces red snapper fillets, cubed
12 ounces cod fillets, cubed
1 dozen shrimp, peeled and deveined
1 tablespoon anise-flavored liqueur
½ teaspoon freshly ground pepper

Heat oil in a Dutch oven over medium heat. Stir in onions, leek, and fennel; cook, stirring frequently, until translucent, about 10 minutes. Stir in garlic, orange peel, thyme, and saffron; cook 1 minute. Add tomatoes, clam juice, water, wine, and salt. Bring to boil. Reduce heat and briskly simmer uncovered 30 minutes.

Increase heat to high and stir in clams. Cover and cook just until clams open, about 5 minutes. (Discard any unopened clams.) Stir in fish, shrimp, liqueur, and pepper. Simmer covered just until fish is cooked, about 5 minutes more. Ladle into bowls. Makes 6 servings.

---

garlic, thyme, and saffron and add them together with the orange peel to the sauté, stirring thoroughly. He would then go to the stockroom and issue out the tomatoes, clam juice, and wine. He would go to the water tap and issue the water, adding those ingredients into the sauté while salting to taste. Adjusting the flame to a controlled simmer, he would return to the refrigerator and issue the shrimp and clams, peel and devein the shrimp and scrub the clams, and return them to the fish compartment. He would follow the same procedure with the fish. He would then reissue the clams and add them to the broth. Five minutes later, he would issue the chopped fish and peeled and cracked shellfish, adding them to the broth along with the anise-flavored liqueur and ground pepper. Covering the pot, he would let it simmer for 5 more minutes. Voilà! A finished product ready for shipping to the customer. Needless to say, that would be a rather complicated way to prepare bouillabaisse.

Defining Levels in the Bills of Material and Routings 49

The point of this culinary exercise is to demonstrate the problems that can arise with an overly structured manufacturing process and its impact on the data foundations. Several levels of bill structure, and lengthy routings, and their associated maintenance are required. The reporting and move transactions can overwhelm the process. How, then, do we go about this process of defining the structure of the manufacturing process?

As discussed in Chapter 2, the manufacturing process is defined in terms of material content (bill of material) and the required manufacturing steps (routing). The bill of material *structure* has two major elements. The first is architecture, or how the bill of material is organized, which will be discussed in Chapter 6. The second element is the depth of the bill, or how many levels it contains. Levels are determined by interruptions in factory flow or by stocking materials at intermediate points in the production process.

An example of a manufacturing process is shown in Figure 3-2. Our

Figure 3-2
Manufacturing Process

Level	Item	Operations	Steps
Level 0	Finished Product		
		Op. 30 / Op. 20 / Op. 10	3 Steps
Level 1	D		
		Op. 30 / Op. 20 / Op. 10	3 Steps
Level 2	C		
		Op. 40 / Op. 30 / Op. 20 / Op. 10	4 Steps
Level 3	B		
		Op. 20 / Op. 10	2 Steps
Level 4	A		
		Op. 30 / Op. 20 / Op. 10	3 Steps
Level 5	Raw Material		

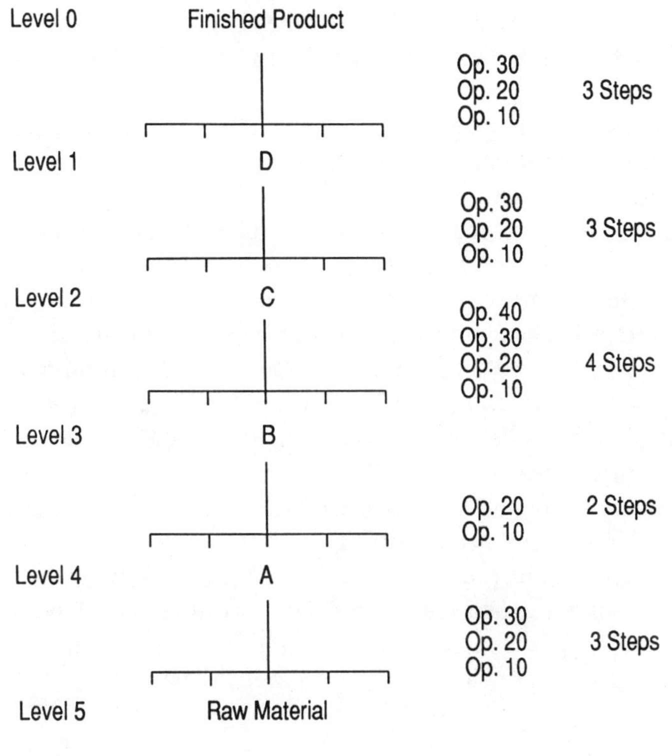

example has six levels and 15 routing steps. The question is, Why are there six levels and 15 steps, or why have we identified this information in our data foundations? A level in the bill of material is usually assigned because we stock or store this material, create plans and schedules for this material by item number, or build information about this item, such as its cost, its lead time, or its order policy at this stage of manufacture. The direction of manufacturing companies today, however, is not to stock parts, create work orders, or maintain the associated reporting. The goal is *flatter*, or more shallow, bills of material, because production flows smoothly with fewer interruptions.

The same is true about routing steps. Routing steps are assigned so that we can specify the manufacturing process activities and the sequence of those activities. They are also established to set standard times for setups and work, to perform detail capacity planning, to track where work is as it flows through the plant, to schedule operation-by-operation work flows, and to monitor queues. These functions are necessary in order to measure our work activity and costs against standards. Here again, the direction of manufacturing companies today is to eliminate queues, reduce setups, end costly reporting, and not create a lot of data maintenance. And the goal is flatter, or more shallow, routings based on production flow utilizing cells, flexible labor, and kanban scheduling.

In today's environment, companies want to receive raw materials and purchased parts on day 1 and use the items and ship the finished product on day 2. Our structure example might then look like Figure 3-3. This has led to significant reductions in wasteful data maintenance and large, cumbersome data foundations. But data foundations do not drive the manufacturing process. They *represent* the manufacturing process. Therefore, to reduce information maintenance with flatter bills of material and shallower routings, the basic way the product is produced needs to be restructured. In this book, we cannot tell you specifically how to do this, but we can suggest a process that allows a company to address its structure opportunity.

Each company has its own products, processes, marketplace—even its own rhythm and personality. Therefore, a company's assessment of how to structure its products should be done internally by the company staff. This is not an assignment that should be hired out. How a company structures its products has a far-reaching impact, and implementing a successful structure change requires company-wide understanding and

Defining Levels in the Bills of Material and Routings   51

Figure 3-3
Manufacturing Process

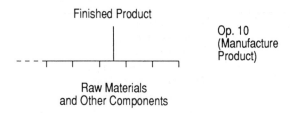

support. Let us now address restructuring, using the CCS Carpet Company as an example.

The CCS Carpet Cleaning Company decided to reevaluate the manufacturing process of one of its newer products—the manual version of the standard carpet cleaner, marketed as the Spotbuster. Management was very excited about the idea because growth in this product line was forecast, and there was a considerable opportunity to streamline the process. There was also a plan to implement MRP II systems in the Spotbuster Division. Getting the data foundations right was appropriate preparation for MRP II.

The first step was to build a team of people representing every functional area in the business. It was extremely important that every function be represented. All organizations needed to define their data foundation needs, their data maintenance responsibilities, and their requirements to support data accuracy. Participation was also necessary for company-wide support of the change. When one of the CCS staff asked who did not need to be represented, the general manager answered, "Perhaps the lawn maintenance group and the cafeteria staff, but I can't think of anyone else."

A *data foundations task team* was established, its members coming from the various departments listed in Figure 3-4. This team was responsible for structuring not only the bill of material but also the associated routings and item master data files. The team members were instructed, however, to create this data foundation around the simplest possible manufacturing process and flow. Management's primary directive was to develop the best possible factory flow and hence eliminate all wasteful and unnecessary activity.

CCS management recognized that team education was necessary and

Figure 3-4
Manufacturing Data Foundations Task Team

Design Engineering
Manufacturing Engineering
Sales and Marketing
Sales Order Processing
Planning and Scheduling
Production Operations
Finance
Quality Assurance
Service
Information Systems

---

sent the task team to classes on manufacturing flows, Just-in-Time, and structuring data bases. Soon after completing these courses, the team began to meet regularly in a series of business meetings to discuss what was learned, how it applied to CCS, and where the opportunities were to implement management's objectives.

One of the first steps the team took was to identify the current structure of the Spotbuster product. This wasn't difficult. Only raw materials and the final product were identified by part number, and bills of material and routings were sketchy at best. The division basically ran on product knowledge and people guesswork. Late shipments, high inventories, and part shortages were all symptoms of the lack of solid data foundations and good planning and control systems.

A second important step was to identify each function's requirements for information to ensure that everyone's needs were considered. Input was collected, discussed, and rechecked to make sure the requirements were fully understood. Many of these requirements changed as various alternatives were identified. But this was easier to accomplish once the original requirements were clearly understood.

The third step the team undertook was to identify alternative structure levels. Four levels were initially selected, and a fifth was added before the analysis was completed.

To identify transaction or control activity associated with each alternative, the team established codes that signified each activity type and posted these codes to each alternative structure. Here are the codes they selected:

Defining Levels in the Bills of Material and Routings

WO—Work order required

PO —Purchase order required

SI —Stockroom issue

SR —Stockroom receipt

SP —Inventory stock point

PR —Process/routing required

RS —Routing steps/operations

Although these activities were basically accepted and expected in some companies, the CCS team worked diligently to challenge their necessity, attempting whenever possible to streamline reporting, data base maintenance, and any non-value-adding activity. The team then summarized the advantages and disadvantages of each alternative structure.

Each approach was summarized and posted to a matrix, as illustrated in Figure 3-5. The numbers resulted from the following alternative approaches:

Option 1. Every manufacturing step was assigned a part number; every routing had a single operation.

Option 2. Status quo. Only purchased material and the final product were assigned a part number. There were no routings. This option required the least reporting, but the CCS team realized that some level of planning and reporting was required to gain control.

Option 3. Ten subassembly structures were identified as a minimum level of data required to support control. Each operation of Option 1 was attached to the appropriate subassembly routing.

Option 4. Same as Option 3, except a thorough review of the manufacturing steps identified only fifty necessary routing operations.

Option 5. A final review allowed three subassembly, three routings, and seven operation steps to be eliminated. This option, however, caused some concern for the production department and the accounting group.

**Figure 3-5
Statistical Summary**

OPTION	PART NUMBERS	MPS ITEMS	WORK ORDERS	PURCHASE ORDERS	STOCKROOM ISSUES	STOCKROOM RECEIPTS	PROCESSES	ROUTING STEPS
1	206	1	160	46	205	160	160	160
2	47	1	1	46	46	1	1	0
3	57	1	10	46	55	10	10	160
4	55	1	10	46	55	10	10	50
5	53	1	7	46	52	7	7	43

The team met frequently with middle management and the department heads that were responsible for Spotbuster operations. Their feedback was helpful and encouraging. A number of important points were presented that affected various alternatives, and when the analysis was completed, the team had full middle-management support.

A subsequent meeting was held with the senior management responsible for the product line. Two proposals, Options 4 and 5, were presented. At first, the differences seemed negligible, but the task team leader carefully explained the arguments supporting each alternative. Eventually, Option 4 was selected, because it provided for a little more control. It was also agreed that once the new systems were implemented, another look at a move to the less wasteful Option 5 approach would be initiated. This, then, became the foundation for the way CCS planned to manufacture and control the new Spotbuster.

One of the main reasons for the success of this project was that the requirements of every functional area in the factory were addressed. Every attempt was made to facilitate a flow manufacturing approach and eliminate waste whenever it was encountered. Because the manufacturing flows were well defined and the necessary data foundations determined, the implementation of the planning, scheduling, and control systems were a real success. Most important, everyone in the organization understood the approach and supported good data control. The Spotbuster group continues to report very high levels of data accuracy.

The success of the Spotbuster program at the CCS Carpet Cleaning Company was so impressive that another team was formed to research older product designs. Since the process worked well with a relatively new product, it was hoped that they could translate what they had learned to other carpet cleaner designs. As expected, they were able to redesign processes, flatten bills of material and routing structures, and simplify their data foundations on their entire line of carpet cleaners. The time and money saved were substantial.

An example of this effort can be seen with the #0123 CCS carpet cleaner. A final assembly line was set up with a feeder station that supplied tank assemblies at the rate of final production. Also, they only bottled enough CCS solution for the next day's run. The tank assembly and the customer pack were now assembled on the final line.

Similarly, CCS analyzed its chemical plant processes. Because of the nature of process manufacturing, where bills of material are already fairly flat, it was originally thought that little could be done. They were wrong. With a great deal of excellent input from the mixing department and a little rethinking on the part of the chemistry group, it was discovered that bulk CCS cleaner could be manufactured in a single process combining both dry and liquid components. This effectively eliminated a level in the process bill. It also significantly reduced storage space and intermediate inventory cost while solving a moisture problem they were having when they stored #4424 Hi-Glo compound.

CCS's new, flattened bill of material was a single-level structure in which the tank disappeared. They still had a #6221 bill of material to bottle CCS concentrate, but the #3427 CCS concentrate bill of material was now also single level (see Figure 3-6).

CCS's efforts in this area eventually involved the tank parts production group. They were already operating with a single-level bill of material structure, but wanted to attack their multistep routing. After a lot of discussion, again including the production personnel, it was decided that a kanban approach (replenishment triggered by completion) could be accomplished. It was also determined that with so little in process and for such a short time, labor and shop floor control reporting added no value to the product and was not needed. With some minor machine movements, a cell was created that allowed the plant to fabricate tank parts completely, move them to paint, and then move them again to the final line. The tank parts moved on carts from department to department. The carts were the kanbans. An empty cart returned was authorization to build more. The entire process took two hours. The routing was reduced from nine steps to two, with no production reporting other than daily production completions. Once in place, the production team increased its efforts on improving quality, which was already significantly improved, and on further shortening the two-hour throughput time (see Figure 3-7).

MAKING IT HAPPEN

The process described in this chapter for defining the bill of material structure is something that every company must do. It is not, however, a process that can be done for you. It is best done by those who are involved in the process and who understand company procedures, the intended marketplace, and the way the product goes together. Doing it yourself is also a vital step to ensure ownership. But doing it yourself does require that those involved be well educated on the issues. Some professional guidance to assist in structuring the necessary processes or

Defining Levels in the Bills of Material and Routings 57

**Figure 3-6
"Flattened" Bill of Material**

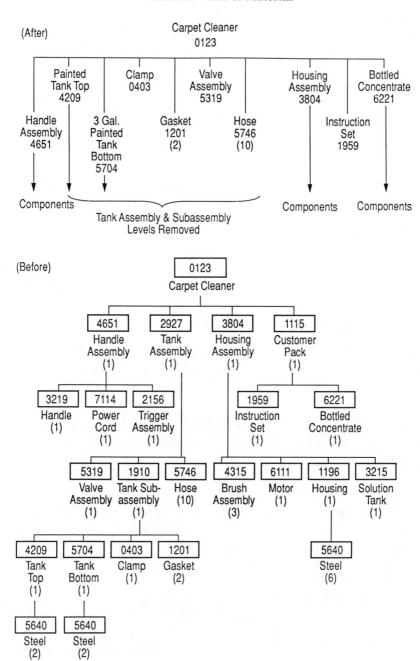

Figure 3-7
"Flattened" Process Routing

(Before)

Parent: 5704
Description: Tank Bottom

OP. NO.	DEPT.	W.C.	DESC.	SU.	RUN	LABOR	TOOL
10	FAB.	120	Cut Tank Blank	2.0	.010	16	1824
20	MACH.	240	Drill Hose Inlet	1.0	.020	13	—
30	FAB.	160	Form Gasket Ring	2.0	.010	16	1893
40	FAB.	160	Form Box	2.0	.010	16	1762
50	WELD.	370	Weld Box	0	.030	18	—
60	FAB.	110	Deburr as Necessary	0	.005	04	—
70	INSP.	490	Inspection-Instruction 10460	2.0	0	11	—
80	PNT.	560	Paint	0	.020	10	—
90	INSP.	410	Inspection-Instruction 10390	0	.002	14	303

(After)

OP. NO.	DEPT.	W.C.	DESC.	SU.	RUN	LABOR	TOOL
10	FAB.	165	Form Complete	0.2	.030	15	—
20	PNT.	560	Paint	0	.020	10	—

to provide insight into various methods of handling data foundation structure can be helpful. However, the task itself must be a company endeavor.

The process of defining and structuring the manufacturing data foundations requires a team effort. Invariably, there are some difficult issues that will arise as you go through this structuring process. The only way to deal with these issues is to sit down around a table and sort them out. Both the benefits and the disadvantages of the potential approaches must be identified and discussed.

It is generally advisable to use a representative product, one that depicts the way you manufacture. This initial product structure will normally establish a format to structure your products. If such a product does not exist, consider using a *dummy* product that can represent all

Defining Levels in the Bills of Material and Routings 59

your manufacturing processes. Use just enough item numbers to illustrate the different steps of the process.

It is also important to realize that this effort can't simply consider bills of material. It must also embrace product routings, and you can't create routings without understanding how the shop floor is run and how things are actually put together. This is why the input of manufacturing operations is always required, to describe how these products must go together and at what levels information should be reported. Note also that all costing and accounting requirements must be considered.

The final aspect of this process is to approach it from a JIT perspective, where less reporting is always best. The ultimate objective here is to eliminate all the waste in the system involving work orders, purchase orders, file maintenance activities, part numbers, and stocking transactions. It was not until the CCS Carpet Cleaning Company took hold of these concepts and became a company dedicated to eliminating waste that it was truly prepared to deal competitively through the implementation of improved manufacturing flow and data foundation simplification.

Chapter 4
Achieving Accuracy and Completeness

One day, one of the managers of the CCS Carpet Cleaning Company was building a model airplane with his daughter. After placing the final decal on the wings and touching up the paint on the plane's belly, they discovered, to their horror, one final circular part—#73—in the box. They searched the drawing that had come with the instructions, but they couldn't find #73 listed anywhere. They looked at each other. The plane certainly appeared to be complete without part #73. So they conspiratorially tossed the *extra* part, the box, and the instructions away.

Later that evening, however, the manager began to think about that unused part. Had that "extra" part been a piece from the plane he was supposed to fly on the next day, he certainly would want to know that it was in the right place, performing its proper function. The point came up again when he opened a bottle of medication. What if the pharmaceutical company that produced the pill had omitted an ingredient or put too much of something in? The consequences were unthinkable.

He realized, of course, that the same standard for accuracy and completeness that he demanded from the companies he relied on in his daily life was what he should expect from the CCS Carpet Cleaning Company. Where would CCS be if the solution tank was inadvertently left off the bill of material? They'd be wiping up a good deal of cleaning solution. Or suppose that the quantity per on the brush assemblies had been registered as two instead of three? In either case, the company would have a major problem when it came to planning, scheduling, costing, and building their carpet cleaners, and those cleaners certainly wouldn't be products they would want to ship to their customers.

BILL OF MATERIAL ACCURACY

Bills of material must be 98 to 100 percent accurate. This means that at minimum, 98 percent of all single-level structures on file must be 100 percent correct. Experience teaches that material requirements planning with bills at less than 98 percent accuracy simply fails. So many schedules and forward plans are incorrect that users can't trust the information. Exception messages grow until it becomes impossible to gain control before the next set of orders and messages are generated. The financial outlook becomes suspect, and second-guessing becomes the common practice. Our target bill of material accuracy is 100 percent.

Bill of material accuracy is measured at each single level in a multilevel structure. In Chapter 2, we learned that by getting single levels correctly identified in our data files, the computer can retrieve any necessary multilevel information. For a bill of material to be considered correct, the parent item must exist and have proper authorizing documentation (engineering order, specification, etc.) and be identified in the company computer file. The correct components must be listed with the correct quantity of usage (quantity per) for that parent relationship. The unit of measure for each component, though usually not maintained on the bill of material itself, must also be correctly identified. In all cases, the authorizing documentation, the company data base, and the way the product is actually produced must agree (see Figure 4-1).

The audit of a bill of material results in a pass/fail or hit/miss (see Figure 4-2). Whether there is one error or several errors, a bill of material is scored as a single miss. The formula for calculating bill of material accuracy is:

$$\frac{\text{HITS}}{\text{AUDITS}} \times 100 = \text{ACCURACY PERCENT}$$

Two areas of caution need to be mentioned. The first is the *as required* quantity per. Although this might be a nice way to reference an item on the bill of material, it must be recognized that this item will not be planned, scheduled, or costed. The computer kind of chuckles to

Figure 4-1
Bill of Material Errors

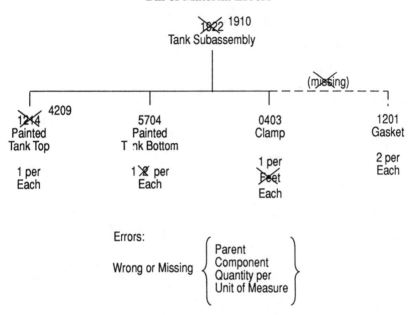

itself when it sees "as required" saying, in effect, "If you don't care, neither do I, so I'll simply ignore the whole thing!"

The second caution regards the use of more than one unit of measure for an item or part number. Although it is sometimes necessary to convert quantities for stores, picking, operator instruction, or supplier ordering, the internal system should operate with a single unit of purchase, unit of receipt, unit of storage, and unit of issue. This is not only a computer planning and costing issue but also a real communication problem. More than once, premiums were paid to rush in kilograms of material when only a few grams were critically short. Conversion tables, though seemingly a practical solution, are often difficult to maintain and sometimes require some complex translation. One frustrated data administrator suggested *beats me!* as the unit of measure for a component that was received in tons, stocked in pounds, and issued in square inches.

This facet of unit of measure differs from the need to maintain *dual* units of measure on an item or routing step. For example, a brass rod producer will track both units and weight produced when extruding.

64 Manufacturing Data Structures

**Figure 4-2
Measuring Bill of Material Accuracy**

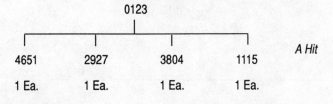

A Hit

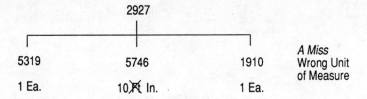

*A Miss
Wrong Unit
of Measure*

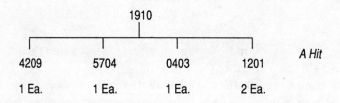

A Hit

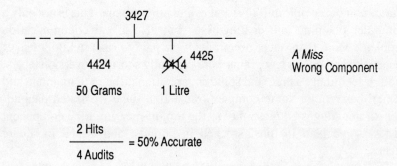

*A Miss
Wrong Component*

$$\frac{2 \text{ Hits}}{4 \text{ Audits}} = 50\% \text{ Accurate}$$

Similarly, trim loss when slitting master rolls of an imaging material to size would be recorded by weight, while slit rolls are tracked by both length and weight.

BILL OF MATERIAL COMPLETENESS

For a bill of material to be complete, it should contain all the items to be planned using material requirements planning and to be included in the calculation of product cost. "All the items" usually includes packaging, consumables, hardware, and raw materials.

By using material requirements to plan these materials, requirements over the full planning horizon are available for supplier negotiation, and plans are responsive to product mix changes. Costs are more accurate, and the material content is precisely described.

Actual order release and issues of these materials may be achieved with simpler minimum/maximum or bulk issue procedures. In some instances, these materials become part of line stocks or are often simply issued as free stock and expensed as overhead material cost. By using coding available in software packages, it is possible to select any number of options in dealing with bill of material components, such as "Issue" or "Do not issue" on a pick list or "Cost" or "Do not cost" in a product cost roll-up.

Typically, such items as packaging, consumables, hardware, and raw materials are identified when product design occurs, since their requirements need to be determined for a product to be produced and shipped to the marketplace. Once loaded onto the data files, engineering change is usually minimal. If changes do occur more frequently, the need to record, plan, and cost such materials is an even more important requirement. For most companies, the policy is simple: If it is part of the material content, put it on the bill. At one company, an item-by-item justification is required to exclude an item from the bill of material. The resulting list of exclusions is reviewed periodically to attempt to reduce the number of exclusions.

Completeness is not in conflict with elimination of waste. A complete definition of material content does not mean more levels in the structure; nor does it mean that replenishment techniques like kanban cannot apply. It does mean that the definition of the manufacturing process contains a complete identification of material content that can then support whatever planning and costing mechanisms the company chooses to implement.

AUDITING BILLS OF MATERIAL

There are four primary ways to audit the accuracy and completeness of bills of material. One approach is to perform a conference room audit with a bill of material audit team. This team would normally include representatives from development or design engineering, finance, manufacturing, operations planning, quality control, and manufacturing engineering. This team audits samples of the single-level bills of material representing finished products and partially finished products such as subassemblies or intermediates. Companies will often do a Pareto analysis of items, which ranks them in order of descending usage and chooses the more active items for initial auditing.

After a thorough review of the bills to check that part numbering, parent/component relationship, quantity per, and unit of measure agree with the authorizing documentation (engineering order, specification, etc.), the team should decide if each bill actually represents the way the product is put together. This almost always involves the help of supervisors and production workers. A conference room audit is an excellent approach for assessing bill of material accuracy if no other audits or performance measures are currently in place.

A second method for checking accuracy and completeness involves sampling work orders against the bill of material on file as work orders are released. The pick list normally is used as the focus of this audit approach. Material is issued exactly as called for on the pick list, and the audit team, using authorizing documentation and supervisor/worker knowledge, determines if the material issued supports the correct parent-level material content. This approach concentrates on what is actually being manufactured, but could be difficult to track through all the possible manufacturing steps and flow.

A third approach is to allow a parent-level item to be manufactured and then disassemble it to determine if all the correct components, in their correct usage, were used in the production process. One advantage to this approach is that it can be highly visible and gain considerable attention. One disadvantage with this approach is practicality. Disassembling a 747 airplane, a chemical reaction, or a facial powder doesn't make a lot of sense, and it might be an impossible task.

One heavy-truck company used a variation on the disassembly audit approach. The vehicle was not actually disassembled, but it was thoroughly checked for material content. A large, white rectangle was painted on the factory floor near the last assembly operation. Vehicles were randomly selected from final line schedules and moved to the audit area. Results of the audits were posted for all to see, and errors were traced to the source in order to correct bill of material or procedural problems.

The fourth audit approach is to allow the company's greatest experts in producing a product—the supervisors and production personnel—to be responsible for the audit activity. Whenever wrong or insufficient materials are delivered to the work area, production people there report the problem to their supervisor as a product or process discrepancy report. After correcting the problem, the supervisor checks to verify that the problem was not caused by a material handling error. If it was not, the supervisor reports the problem to manufacturing engineering for reconciliation.

For this approach to work, some important support activities need to be in place. When errors are identified, they need to be corrected promptly. Nothing destroys the credibility of the process more quickly than to have a problem reported as corrected and then have it recur. If a material discrepancy is reported and the next material issue isn't correct, the enthusiasm to report it again weakens. If slow response is the rule, floor feedback will stop.

Another company, as an ongoing incentive, pays money for every bill of material error found. This expense is then charged to the department responsible for the error. Most companies, however, don't use financial awards to eliminate bill errors. They rely on education and teamwork to achieve the highest levels of data integrity. There is something in it for everyone.

ROUTING ACCURACY

Routings must be 95 to 100 percent accurate. This means, at minimum, that 95 percent of routings must be 100 percent correct. Once again, if routings are more than 5 percent in error, capacity planning, detail capacity tracking, shop floor scheduling, and work-in-process accounting will be too inaccurate for users of the system to trust the reported information. This leads to second-guessing, a return to informal systems, and the loss of the benefits of good plant scheduling and accounting.

As with bills of material, accuracy is measured at each parent level;

that is, all operations on the routing must be correct for the routing to be correct. To be considered correct, the routing must identify all operations in their correct sequence. All work centers must be correctly stated and times or standards set for each operation. Most companies accept time standards that are reasonably correct and consistent for items flowing through a work center. Standard times of plus or minus 20 percent are acceptable in many companies.

Routing accuracy is also measured on a pass/fail or hit/miss basis, with one or more errors on a routing making it one wrong routing. The same accuracy formula used in bill of material accuracy is used to measure routing accuracy:

$$\frac{\text{HITS}}{\text{AUDITS}} \times 100 = \text{ACCURACY PERCENT}$$

Some companies include labor grades or codes on their routings to plan labor requirements and machine requirements, and tool identifiers to plan nonconsumable tooling requirements and schedules. When this is the case, such companies will generally include the labor and tool identifications as elements to be audited for routing accuracy.

AUDITING ROUTINGS

There are three common ways to audit routings. The first approach is a conference room audit with a routing audit team. This team normally includes representatives from accounting, manufacturing, planning, quality control, and manufacturing engineering. As with bill of material accuracy, production people and supervisors need to be involved. Sample sets of routings are audited, making sure that the sample represents the various active processes operating in the plant. A thorough review of operation identification and sequence, work center identification, and assignment of standards is followed by a check on how the production process is actually performed.

A second audit approach requires the audit team to actually track the production process and test it against the documentation and routings. The problem with this approach is that there may be many lengthy operations, resulting in a long audit process, or there may be many diverse operations, making the process difficult to track. A variation on this method is to check any variances in reported routing activity against the planned routing activity. This method is easier but occurs after the fact, and the cause of errors may be difficult to capture.

The third approach is to let the factory staff conduct the audits. As with the similar bill of material audit approach, the world's greatest experts are in control. Shop floor reporting tells the supervisor when an order is in a particular work center. If the order isn't physically there, a routing error could be the cause. Likewise, orders in the area that are not reported as being there could be another indication of a routing error. Material move errors can also cause such problems and need to be investigated.

The CCS Carpet Cleaner Company audits its bills of materials and routings using a single audit feedback form. One side of the form is for bill of material errors, while the other side is for routing errors. If a supervisor identifies a problem in any one of the four main bill of material issues—incorrect parent item number, parent/component relationship, quantity per, or unit of measure—the form is sent to the bill of material administrator. If an error is found in the routing because an operation, its sequence, or work center is incorrect, or because a standard is way off the mark, the form is sent to manufacturing engineering.

Attached to this form is a standard reply section, where the individual in charge can inform the initiator of the corrective action being taken.

ITEM MASTER AND WORK CENTER MASTER DATA ACCURACY

As with bills of material and routings, a high level of accuracy is required on item master and work center master data files. This is one area, however, where the data requirements do not necessarily demand the same degree of precision as routings and bills. In some ways, item master information is more forgiving data. Lead times, for example, are based on average lot sizes, current supply situations, and a number of other variables, and may be different for each order placed. Times on the work center master may change continuously because of real capacity available.

Data on both master files can typically be identified as either nonmandatory (merely informational) or mandatory. Mandatory item data include those data elements that support the planning and costing systems. Such mandatory data should never be missing from the data files and should be reviewed periodically to ensure that reported information passes basic, reasonable checks.

Most companies find it helpful to perform periodic logical edits of item and work center master information. For example, even though computer logic is incorporated in maintenance routines, it may be useful to identify, for example, items with lead times exceeding a certain threshold, missing codes, or invalid account numbers.

DATA MANAGEMENT

One of the central issues in data accuracy is that of responsibility for maintaining the data files. This can become a debate over centralized versus decentralized control. Our experience is that both approaches work. But a formalized process must be installed. Data responsibility should be clearly assigned, and targets for accuracy should be identified and measured. Everyone needs to understand and support the process and own the data. As usual, people are the key to a successful operation.

In the centralized approach, the data files are maintained by a designated group often called *data administration.* All input flows to this organization, where it is usually verified and loaded into the company data files. Because of the integrated nature of the information, the data administration function also ensures that all data on file are maintained whenever an engineering or specification change occurs.

In the decentralized approach, the data files are maintained at the source. Each organization responsible for a data decision is also responsible for data maintenance. This approach requires that each organization understands the need for data integrity and carries out its data maintenance responsibilities in a timely, accurate manner. To be effective, company management must hold each organization responsible for data foundation maintenance and accuracy as for any other assigned responsibilities.

In both approaches, edit and other errors need to be quickly corrected and reentered into the company data files. Formal procedures should be implemented to make sure that the rapid correction of systems errors takes place, especially in batch computing environments. This activity should also be measured and reported. Most companies will typically track the number of errors by type, the number corrected within the time allowed, and the aging of outstanding errors.

In the centralized approach, error correction often has to recirculate to the source for reconciliation. In the decentralized approach, the individual who is entering the data is often the one who makes the correction.

Achieving Accuracy and Completeness 71

MAKING IT HAPPEN

Everyone has heard the expression, *Garbage in, garbage out*. The truth is, if we don't put the correct data into the system, we won't get the correct data out. The authors consider the accuracy and completeness of a company's data foundations to be a quality issue. The objective is zero defects. The cost of poor quality is high:

- Duplicate data files (no one trusts the other numbers)
- Continuous data debate
- Bad decisions based on bad data
- Inability to make decisions because of poor information or lack of information
- Poor planning
- Poor execution
- Lack of credibility and return to the informal system

Making it happen requires a total company commitment to high levels of data integrity in the data foundations. To determine where a company stands in relationship to this objective, it is often necessary to test the existing data accuracy and completeness. It is also important to identify the major problem areas and allocate the appropriate resources to correct the problems, and bring the data foundations to their desired integrity targets. The initial testing is normally accomplished by use of the conference room audit described in this chapter, making sure that a sufficient sample size is audited to build credibility in the audit result. The next step is the education of everyone involved in the process, building ownership and support of the data accuracy and completeness objective. The education phase is followed by a training period, dealing with all the new data management practices required to achieve and maintain the necessary data integrity levels.

Once the data foundations are at their desired integrity levels, ongoing education and a formal maintenance process (see Chapter 8) provide the means by which a company can maintain highly accurate and complete data files. No one knowingly loads bad data into a system or allows bad data to remain in the system once it has been identified. Continuing audits provide the process control to monitor and measure the data management process. If the process is breaking down, corrective action must be initiated before data integrity goes out of control.

The objectives are high—zero data defects, or 100 percent accuracy. The boundaries are tight—98 percent on bills of material and 95 percent on routings. But the mission is critical. Customer service, reduced cost, and real productivity will not happen without this level of data control. Data accuracy is the cement that holds the data foundation together, and that foundation supports the business objective of competitive success.

Chapter 5

Planning Scheduling and Controlling the Plant Using the Data Foundation

Once the CCS Carpet Cleaner Company accurately and completely defined its manufacturing process in its data files, it began using the time-phased planning capabilities of material requirements planning (MRP) and capacity requirements planning (CRP). These computer programs are based on the universal manufacturing equation that simply outlines how to plan what you need and when you need it by answering these three questions:

1. What are we going to make?
2. What does it take to make it?
3. What do we have?

For material planning, the material requirements planning system gets its demand ("going to make") from a master schedule or from individual independent demand. The bill of material provides the material content of our demand ("take to make it"), and the on-hand/on-order inventory record provides the component availability ("what do we have"). The material requirements process simply, repetitiously tests these conditions as it works its way down through every level of material content for the products a company wishes to produce.

To get an idea of how this process works, let's assume that the CCS Carpet Cleaner Company wants to ship twelve carpet cleaners, and it has six in inventory. The system would quickly determine that six are available and that there is a need to manufacture six. The computer program would then go to the bill of material and identify all the components required to assemble a carpet cleaner and create a demand for six complete sets. Each component would then be checked for its inventory availability, either on hand or on order. If sufficient inventory was lacking, demand for additional components would be created to make available sets of parts for six carpet cleaners. This material requirements planning approach is called *dependent demand planning*. Demand for component items, defined by the bill of material, *depends* on the number of parent items the company needs to make.

Once the material requirements—both purchased and manufactured—are identified through the use of material requirements planning, many companies need to determine the availability of detail capacity with which to meet the production requirements. This is done through a computer system known as capacity requirements planning, in which the universal manufacturing equation is applied to determine capacity constraints.

Demand for capacity comes from master scheduling and the material requirements planning process ("need to make"). The definition of capacity required comes from the routing files ("take to make it"); the planned and demonstrated capacities are found on the work center master file ("what do we have"). The capacity requirements planning process takes the projected replenishment for an item from MRP and, using its routing file, calculates required capacity for each work center. This need, along with all the other calculated needs for a given work center in a given time period, is compared to demonstrated and planned capacities and out-of-balance conditions (see Figure 5-1).

MATERIAL REQUIREMENTS PLANNING

Because material planning is such an important process for manufacturing companies, an understanding of how MRP uses the bill of material and the lead times recorded on the item master file is appropriate.

Figure 5-1
Universal Manufacturing Equation

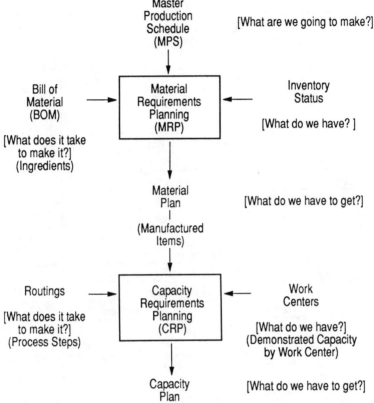

Turning the bill of material sideways is an easy way to visualize the lead times for component items (see Figure 5-2). The *cumulative lead time* for producing a cleaner is fifteen weeks. Cumulative lead times, also called stacked or end-to-end lead times, establish the longest sequence(s) of single-level lead time items required to manufacture an item. It's similar to the critical path through a project planning network. If CCS knows the dates on which they need carpet cleaners, time-phased material requirements planning will tell them when they need the matched sets of parts at all levels to satisfy that need. Since they allow one week for the plant to final assemble the cleaner, they need level one items

Figure 5-2
Parallel Lead Times

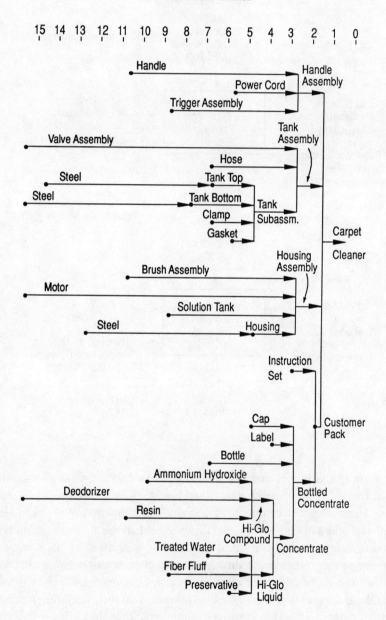

when they start to make the cleaners, one week before they're due. This system not only identifies component requirements but also times them so that the company can have matched sets of components when they need to start the parent. Material requirements planning does this right across the entire planning horizon, through all levels of the bill.

Even though cumulative lead time is fifteen weeks, the planning horizon would go far beyond this time frame. Most companies plan through a horizon of twelve months or more to support supplier negotiations, longer-term capacity planning, and rolling yearly financial outlook (using material and capacity plans as the basis for cost projections).

Now let's look at a detailed material requirements plan for the CCS Carpet Cleaner Company. In Figure 5-3, we see a master production schedule (MPS) that calls for 40 cleaners a week for this month and 45 a week from then on. The planning horizon, of course, would go out much farther than this, but for this example, we'll look at just eight weeks. The master schedule determines what the company is going to build. According to the bill of material, they're going to need one #3804 housing assembly for each cleaner. The first thing they do is offset this requirement by one week: They need #3804 one week before they need cleaners in stock. The demand then is for 40 in weeks 1 through 3, 45 in weeks 4 through 8. The requirements in week 8 come from the week 9 master schedule, which we don't see in this figure. We also no longer see the housing assembly requirements for the 40 cleaners to be produced in week 1, since these components are already issued to work-in-process.

By taking the master schedule demand at level 0 down to the level 1 components and offsetting the demand by one week, CCS can develop *projected gross requirements* for the #3804 housing assembly. This is an example of dependent demand. Planned order releases at a higher level create projected requirements at the next lower level. This process of level-by-level requirements determination is often called *exploding the bill of material* or a *bill of material explosion*. The computer systems establish a direct link between the planning logic and the data base of material content.

At this point, we haven't looked to see if there are any #3804 in stock. Let's see. The computer has projected down the requirements for #3804. CCS currently has 60 housing assemblies in stock. But the system reveals not only what they've got, but also what they're going to get. Scheduled receipts represent future deliveries and future inventory. Such receipts are part of the on-order inventory record and are tied to a

Figure 5-3
Material Requirements Planning

Part No.: 123   Lead Time: 1 Week

MPS			40	40	40	40	45	45	45	45

Part No.: 3804   Lead Time: 2   Order Qty.: 60
Housing Assembly

		1	2	3	4	5	6	7	8
Projected Gross Requirements		40	40	40	45	45	45	45	45
Scheduled Receipts			60						
Projected Available Balance	60	20	40	0	-45 / 15	-30 / 30	-15 / 45	0	-45 / 15
Planned Order Release			60	60	60		60	60	60

Part No.: 1196   Lead Time: 2   Order Qty.: 120
Housing

		1	2	3	4	5	6	7	8
Projected Gross Requirements			60	60	60		60	60	60
Scheduled Receipts				*120					
Projected Available Balance	30	30	-30	30	-30 / 90	90	30	-30 / 90	30
Planned Order Release			120		120				

*Reschedule-in Exception Notice

purchase order, supplier schedule, or work order or build rate plan. In this example, there is a scheduled receipt of 60 in week 2. At this point, the people at CCS know what they need, know what they have, and know what they're scheduled to get. They also know what it takes to make it, because they've just performed the bill of material explosion.

What the MRP process does for CCS is figure that if they have 60 and they're going to use 40 next week, then they're going to have 20 left over at the end of period 1. In period 2, they'll have 20 and they'll be receiving 60, which makes 80, and they'll need 40, which will then leave them with 40 at the end of period 2. In period 3, subtracting the 40 will leave them with 0.

In period 4, they'll need 45, and they project no available inventory to cover this demand. They also have no scheduled receipts that might be rescheduled, so their material requirements planning system creates a planned order release of 60 for period 2. The order quantity of 60 comes from the order policy data on the item master record. The determination that the order needs to be released or started in week 2 comes from the two-week lead time data on the item master record. They'll need to start them in week 2 in order to have them in week 4. Their calculation then assumes that the 60 will be received in period 4, and since they need 45 of the 60 to satisfy demand, their projected available balance will be 15 at the end of the period. Continuing in this manner, material requirements planning projects planned replenishments needed in periods 5 through 8 and across the entire planning horizon.

Proceeding down another level in the CCS bill of material, we reach the #1196 housing. CCS needs one #1196 per each housing assembly. In period 1, there is no demand for these assemblies. In weeks 2, 3, 4, 6, 7, and 8, however, their requirements will be 60 in each week. Again, we see the planned order release at the higher level driving the projected gross requirements at the lower level. An exception occurs with the #1196 housing, however, which will require the planner's action. The 30 housings on hand will not cover the 60 required in period 2. A scheduled receipt of 120 is not due until week 3, and they're projecting a minus 30 in period 2.

Whenever need dates and due dates do not align, material requirements planning issues exception messages to inform the material planners that corrective action is required. In this case, the exception message would recommend moving or rescheduling the order for 120 from period 3 to period 2. The exception message would not recommend placing a new order to cover the minus in this situation, since it is easier to reschedule an existing order first. Computers are programmed to "think" this way.

The eventual decision has to be reached by a person, not by a computer. It may not be possible to bring in the entire order because the company may not have the material or capacity. It also may not be possible to bring in part of the order. It may become necessary to notify a higher-level planner, or even the customer, of a short-term shortage. What we want the computer to do is tell the CCS staff when they have a potential problem. Then people can take the appropriate corrective action.

This planning process is identified as level-by-level time-phased material requirements planning. As noted earlier, it is also known as dependent demand planning, since lower-level item requirements depend on the number of parent items the company plans to produce.

The need for bill of material accuracy when planning material is painfully obvious. As discussed in Chapter 4, parent/component item numbers must be correct, all components must be listed, and quantity per and unit of measure information must be correct. *Reference* or *as-required* quantity per data in the bill of material may be good information for those reading a bill, but it does nothing to support the dependent demand planning process.

This same fundamental time-phased planning also works for CCS chemical products.

The schedule for the #3427 concentrate is increasing over time (see Figure 5-4). Planned order releases occur in weeks 3, 6, and 8. The projected gross requirements for the ingredient #4425 Hi-Glo Liquid thus becomes 4,000 liters in week 3, 6,000 liters in week 6, and 8,000 liters in week 8. Since there is no inventory and there are no scheduled receipts for Hi-Glo Liquid, the material planning logic has created planned order releases one week earlier, considering the one-week process time to manufacture this ingredient.

The level-by-level planning process continues and plans the required amounts of #3811 preservative to support the Hi-Glo requirements. Using the 0.02-liter quantity per recorded on the bill of material (refer back to Figure 2-7), a projected requirement for 80 liters in period 2, 120 liters in period 5, and 160 liters in period 7 is calculated. Since #3811 preservative is purchased in 55-gallon drums, an equivalent amount of 208 liters has been stored in the item master file as a multiple-lot size modifier. After inventory netting, requirements for 208 in period 4 and 208 in period 6 are projected.

The basic material planning logic is universal. A single set of data foundation files supports a valid material planning process for both the carpet cleaner fabrication and assembly plant and the concentrate process plant. CCS is therefore operating both businesses—different in

Figure 5-4
Material Requirements Planning for Process Products

Part No. : __3427__ Lead Time : __1 Week__ Order Qty. : __1000__
Concentrate

Planned Order Release					4000		6000		8000

Part No. : __4425__ Lead Time : __1 Week__ Order Qty. : __1000__ Multiple: __1000__

Hi-Glo Liquid	1	2	3	4	5	6	7	8
Projected Gross Requirements				4000			6000		8000
Scheduled Receipts									
Projected Available Balance	0	0	0	-4000 / 0	0	0	-6000 / 0	0	-8000 / 0
Planned Order Release			4000			6000		8000	

Part No. : __3811__ Lead Time : __1 Week__ Order Qty. : __208__ Multiple: __208__

Preservative	1	2	3	4	5	6	7	8
Projected Gross Requirements		80			120		160	
Scheduled Receipts								
Projected Available Balance	128	128	48	48	48	-72 / 136	136	-24 / 184	184
Planned Order Release					208		208		

many ways—with the same material planning software supported by a common set of total company data files.

CCS has to deal with yield and shrinkage, and the same approach is used at both plant sites. For many items, the problem is handled by loading a shrinkage factor in the item master file. The CCS software system then simply increases the starting quantity on each order to cover the anticipated loss.

In those cases where the loss on an item differs when used on different parents, the problem is handled by a bill of material scrap factor in the software. In this case, component requirements are increased as the dependent planning process calculates each lower-level demand. As an alternative, some companies increase their quantity per amounts.

Yield factors at the item level or in the bill of material record allow CCS to cover any planned yield or shrinkage. These same factors also allow CCS to overissue components and ingredients on pick lists and batch tickets to manufacture required amounts of the parent item.

ALTERNATE AND TEMPORARY SUBSTITUTES

Occasionally, the CCS Carpet Cleaner Company needs to temporarily vary from its standard bills of material for a particular amount of production. Sometimes, they use *alternative* components that are preapproved for occasional use. Once in a while, they use *substitutes*, which are components temporarily approved with engineering deviations. The software allows alternatives to be maintained on the bill of material structures, but they are ignored in the planning and costing process. CCS has learned to maintain strict engineering control over alternatives and substitutions, depending on them only when necessary.

Recently, CCS had an order that contained both an alternative and a substitute. The planning system called for a planned order release of 45 units of #2927 tank assemblies. (This tank assembly is made from a #5319 valve assembly, a #5746 hose, and a #1910 tank subassembly. Figure 5-5 illustrates the planned component requirements generated when a planned order explodes the bill of material file.) The company's planner always reviews these orders early in the planning cycle, since there are sometimes problems with the supply of the #5319 valve assembly. Engineering authorized the use of #5353—an alternative valve assembly—which is a completely interchangeable assembly but more expensive to purchase and slightly less reliable. Needless to say, the company uses these only when it can't get the #5319.

Figure 5-5
Alternates and Temporary Substitutions

Planned Orders Use the Bill of Material

Planned Order
Part No.: 2927
Description: 3-Gal. Tank Assembly

Qty.: 45
Start: 11/1
Complete: 11/8

Planned Requirements

COMPONENT PARTS	DESCRIPTION	QUANTITY PER	U.M.	REQUIRED QUANTITY
1910	3-Gal. Tank Subassembly	1	EA.	45
5319	Valve Assembly	1	EA.	45
5746	Hose	10	IN.	450

Production Scheduler May Use an Authorized Alternate or Temporary Substitute by Creating a Firm Planned Order

Firm Planned Order

FPO: 7762
Part No.: 2927
Description: 3-Gal. Tank Assembly

Qty.: 45
Start: 11/1
Complete: 11/8

Firm Requirements

COMPONENT PARTS	DESCRIPTION	QUANTITY PER	U.M.	REQUIRED QUANTITY
1910	3-Gal. Tank Subassembly	1	EA.	45
5353	Valve Assembly	1	EA.	45
5336	Hose	10	IN.	240
5746	Hose	10	IN.	210

As the planner was reviewing this order, stores called and reported that they had located a box of 24 hoses with item #5336 and couldn't seem to determine any requirement for it. The planner recognized this item as the old hose that was used in the #2927 tank assembly before it was replaced by the #5746 hose. The planner asked for and got a one-time engineering deviation to substitute and use up the old-style hoses.

The planner now needed to do two unusual planning activities. The first was to lock up this nonstandard bill of material configuration for forward planning without actually releasing the order, since its start date was still several weeks out. The second thing the planner wanted to do was not to modify the preferred or standard bill of material. All other orders were to be planned using the standard bill of material record.

To achieve these objectives, the planner created a *firm planned order*, and for this order, set up the one-time component configuration required (see Figure 5-5). At CCS, a firm planned order explodes the bill of material, extending it by the quantity on order, calculates required dates, and automatically posts the resulting requirements to the requirements file. We often refer to the bill structures on the requirements file as *live* structures and to the bills of material on the bill of material file as *master* structures. Usually, the structures are identical. When alternative or temporary substitutions are needed, direct maintenance is performed to the requirements file so that the different configuration is captured.

The terminology relating to the requirements file varies. The requirements file itself is sometimes called the *allocations* or *time-phased allocations* file. It is also known as the *demand* file in a supply/demand sense. A shop order to produce ten of a parent item represents a *supply* of ten, but it creates a *demand* (requirement) for ten sets of component items. The use of the file to modify or override the standard bill of material for a particular order leads to the various expressions *one-time bill of material*, *tying a bill of material to an order*, and *order-specific bill of material*.

It is important that the requirements tied to CCS's firm planned order #7762 be interfaced and used by material requirements planning for component planning. Similarly, when the firm planned order is released, it is important that the pick list for this particular order specify the specially planned components. This is not a problem at CCS. Since all pick lists are actually generated from the requirements file, the planned component requirements are requested (see Figure 5-6).

CAPACITY REQUIREMENTS PLANNING

The universal manufacturing equation mentioned earlier (see Figure 5-1) introduced the detailed capacity planning process, capacity requirements planning, which is tightly linked to MRP. The routing or process information underlies capacity planning in the same way that the bill of material underlies material requirements planning.

Using the routing for an item, capacity requirements planning *explodes* material requirements planning's planned orders for all manufactured items. Detailed scheduling logic assigns operation start and due dates to each operation step in the routing.

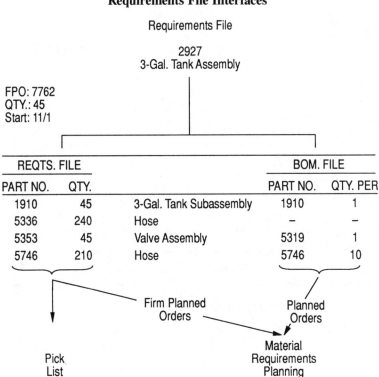

**Figure 5-6
Requirements File Interfaces**

Perhaps the most common logic is a *back-scheduling* routine (see Figure 5-7). Back-scheduling techniques assign a due date to the final operation based on the planned order due date from material requirements planning. Standard setup/changeover and run hours required are calculated by extending the planned order quantity by the standard time factors in the routing. Elapsed time required for the operation is then calculated using shop or workday calendar information, efficiency or utilization factors for the work center, built-in rules for rounding partial shifts or days, and factors allowing for any planned queue times. The end result is a calculated operation start date. Built-in rules allowing for material movement and other activities are then applied to calculate the operation due date of the preceding operation. The routine continues backward in this fashion to the initial operation.

For example, imagine that you are at work and you have a doctor's

Figure 5-7
Back-scheduling an Order

Part No.: 1196 Housing
Qty. 300

Production Order: 8447

Order Due Date: 8/28

OP. SEQ. NO.	DEPT.	WORK CENTER	OPERATION	STANDARDS SETUP TIME (HOURS)	STANDARDS RUN TIME (HRS./PC)	STD. HRS. THIS JOB	OPERATION DUE DATE
10	Fab	107	Form	.5	.010	3.5	8/12
20	Drill	124	Drill	1.5	.030	10.5	8/19
30	Fin.	192	Deburr	3.3	.048	17.7	8/27
40	QC	311	Inspect	2.5	0.000	2.5	8/28

This Is the *Live* Routing for Production Order 8447

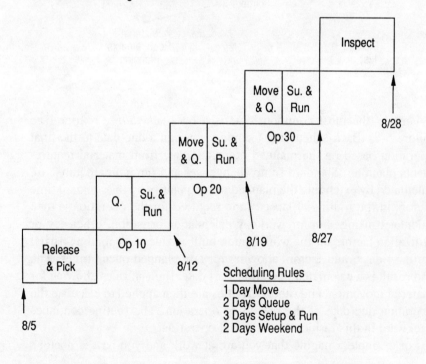

Scheduling Rules
1 Day Move
2 Days Queue
3 Days Setup & Run
2 Days Weekend

appointment at 2 P.M. In trying to determine when to leave, you figure it takes 20 minutes to get from work to the doctor's office, 5 minutes to get to the car, 5 minutes to clean up, and 15 minutes to start the process of getting ready to go. If you were to back-schedule this, it would mean that in order to make your 2 P.M. appointment, you would have to leave the parking lot by 1:40; you'd have to be out the door by 1:35; you'd have to begin cleaning up by 1:30; and you'd have to start getting ready at 1:15.

Other scheduling techniques are also used, such as *forward scheduling* from gateway or bottleneck work centers. The specific logic used is usually a function of the manufacturing environment.

At CCS, back-scheduling logic is used in all areas. This process projects time-phased capacity requirements at each work center for every planned order from material requirements planning. Capacity required for the remaining operations or released shop orders is then included. The total capacity required by a work center is summarized and available for reporting or video display. Bar chart displays are especially convenient.*

Routing data accuracy is a critical factor in determining plant capacity needs, detailed scheduling requirements, and the costs of a company's manufacturing processes. As noted in Chapter 4, all operations need to be captured and correctly sequenced, all work centers identified, and reasonable times established for setup/changeover and run times.

ALTERNATE ROUTINGS AND REWORK OPERATIONS

There are times at CCS when an alternative routing is needed for a particular production order. In one recent situation, an order for 100 #1196 housings needed to be run in a future period at the same time that the high-speed drill machine—needed for operation 20—was scheduled for preventive maintenance. Attempts to reschedule around the planned maintenance were unsuccessful, so production control requested help from industrial engineering to look for alternatives.

When faced with such situations, CCS can sometimes turn to its model shop, and in this instance, industrial engineering was able to set up a single operation process on the model shop's #114 multitool

---

* Additional discussion of capacity planning and scheduling can be found in Appendix A. A more comprehensive treatment can be found in *Gaining Control: Capacity Management and Scheduling*, by James G. Correll and Norris W. Edson.

machine. The shop allocated the time required, and production was authorized to use this alternative process for this particular order of housings.

To realign their capacity requirements properly and to schedule the housings properly, CCS needed to create an alternative routing in its data file (see Figure 5-8). Since all other future orders were to be planned against the established routing, they did not want to change that routing record. Their approach was the same as that used to deal with alternative bills of material. At CCS, the software retains all production information for firm planned and released orders in a file known as the *operations detail file* or *WIP detail file*. It also includes the routing detail. The capacity requirements planning system uses the operations detail file (*live* order process data) to plan and schedule released or firmed-up orders and the routing file (*master* process data) to plan and schedule future or planned orders (see Figure 5-9).

Since CCS did not want to start or release this order yet, they set up a

Figure 5-8
Alternate Routings

Planned Orders Uses the Routing File

Planned Order
Part No.: 1196
Description: Housing

Qty.: 100
Start: 10/4
Complete: 10/18

OP. NO.	DEPT.	W.C.	OPERATION	SETUP	RUN
10	Fab	107	Form on 47L Press	0.5	0.010
20	Drill	124	Drill on Drill Machine	1.5	0.030
30	Fin.	192	Deburr. Labor category 4	3.3	0.048

Production Capacity Planner May Select an Authorized Alternate Routing or a Temporary Routing on the Operation Detail File

FPO: 7781
Part No.: 1196
Description: Housing

Qty.: 100
Start: 10/11
Complete: 10/18

OP. NO.	DEPT.	W.C.	OPERATION	SETUP	RUN
10	Model	120	Form and Drill on Multi-tool 114	0.5	0.010

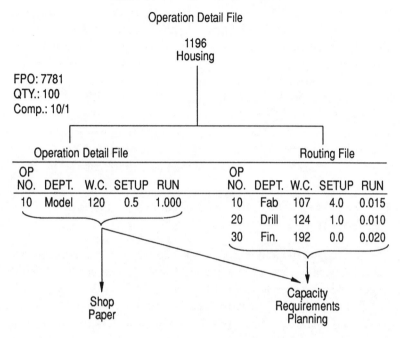

**Figure 5-9
Operation Detail File Interfaces**

firm planned order, which allowed the system to recognize the alternative routing condition without having to release the order. Their computer process initially dated, extended, and copied the standard routing onto the operations detail file. Their planner manually deleted these records and created the alternate record. When the time came to release the order, all paperwork generated by the system reflected this alternate routing process.

At CCS, alternate routings can be maintained on the routing file. These routings are not used to plan future requirements, but can be selected by a special transaction to set up a firm or released order whenever the alternate routing is used. CCS will also set up a routing for any major rework activity on the operations detail file so that it can also be included in their planning and scheduling requirements. Alternate routings are also used with subcontractors. CCS normally does the work in house but can select a subcontractor alternate for a given order or lot.

By tying the production order to the bills of material and routings, the company makes certain that their planning and scheduling are accurate. They plan the item or process that they will use, even when it differs from the preferred. This process ensures that the plan is valid and recognizes the company's real material and process requirements.

CCS has the software capability to tie or specify component and process requirements to an order or schedule when there are exceptions. This is done with both firm planned orders and scheduled receipts utilizing the requirements and operations detail files. The fact that the company can independently maintain these files is very important. Even if they make a change to the standard bill of material, there is no automatic update of the requirements file. It is completely independent. Some software packages simply don't provide these capabilities.

REQUIREMENTS AND OPERATIONS DETAIL FILES: THE *LIVE* FILES

The requirements file and the operations detail file contain material and process information for all firm planned and released orders (scheduled receipts). As discussed, these files allow users to plan and control alternative or nonstandard items and operations on substitute bills of material and routings. These data files, often referred to as the *live* data foundations of a company, serve several other functions (see Figure 5-10).

One such function is to record order status. As an order is released, calculated issue quantities, schedule dates, and standard hours are posted to these files. The extended and dated bill of material is recorded on the requirements file, and the extended and dated routing is recorded on the operations detail file. These mirror the bill and routing from the *master* files unless modified for nonstandard items or processes.

As actual materials are issued to these orders, and as operations are completed against these orders, this information is also posted to these files. Reports and screens are then provided to report the status of material issued or back ordered and the number of operations completed or remaining against an order. Special reporting is often employed to advise of items that are short or past due, and to provide general order status information for the plant and the customer.

Since actual material issues and production (labor/machine) use are reported, these files also serve as the base for actual cost collection. Using material cost data and labor/machine cost data, actual cost systems capture the actual reporting against an order, give the dollar value of the actual activity reported, and establish a material and production expense for each order. Each order is then tracked to a specific customer order or project number, and the total cost of the customer order or project can be developed.

Standard cost systems match the actuals reported against planned

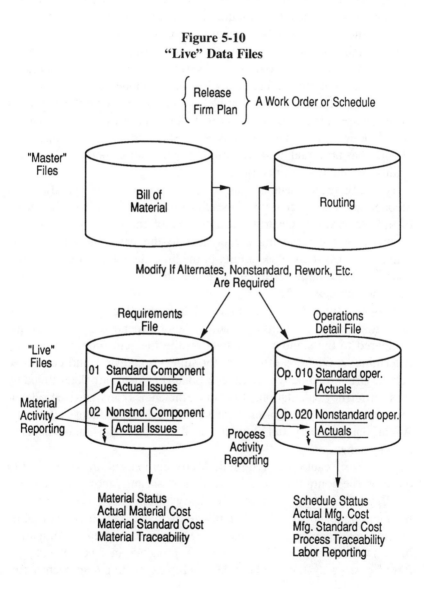

**Figure 5-10
"Live" Data Files**

standard activity and capture variances. This process is based on setting a material standard and a process standard as the planned/expected cost against which the actual is compared. Standard costs are usually established annually but may be revised during the year if a significant swing in cost occurs. For material, the standard is set at the item or part level. For production processes, the standard is based on the routing setup/changeover and run time per piece extended by labor/machine costs and burden rates associated with labor classifications and work centers.

Once the standards are set, material variance is calculated by comparing the standard cost value of the material actually issued to an order against the standard for the item. A variance occurs when components other than the preferred are issued, when components planned are not issued, or when work order completion reporting does not account for the exact use of the components issued. A process variance occurs when actual hours are greater or less than standard, when an operation planned has no actual hours reported, or when actual hours are reported against an unplanned operation.

Typically, these variances are reported to production operations for reconciliation and then booked as a favorable or unfavorable adjustment to inventory value. Clearly, the integrity of the data on file and the accuracy of the reporting are critical to the basic integrity of any standard cost system. The validity of the financial adjustment is a major concern, and a large resource of people and time is often expended to "get the numbers right."

The operations detail file supports labor-reporting systems in a similar way. In these systems, the actual labor expended is reported and compared to the standard hours set on the file. This comparison then determines a worker's efficiency for supervisor review and corrective action. Often, company incentive or bonus payments are determined by these efficiency calculations. Machine efficiencies can likewise be determined. Supervisors are often measured by their department's efficiency. Again, data integrity and reporting accuracy are important elements of such systems.

A word of caution is required. Many companies are beginning to question the continuing role of standard cost- and labor-reporting systems. Such systems are typically costly to operate, tend to measure production personnel rather than utilize their skills and ideas, and often provide management with misleading information. Many companies have, in fact, dropped these costing approaches in favor of *activity-based costing* systems. These less complex costing approaches frequently result in simpler data foundation structures with a significant reduction in data maintenance and factory activity reporting.

The requirements file and operations detail file can also be used to capture traceability data that a company needs to maintain for regulatory requirements, warranty needs, or as protection against possible product liability issues. On the material side, lot number and serial number data are posted to the requirements file for a particular order as part of the inventory transaction that updates the file. Based on the records, material certification data can then be tracked and recovered.

On the manufacturing side, selected process data, such as heat, pressure, speed, acidity, or statistical process control factors, may be added to the operations detail or supporting files as part of the operation completion transaction update. At an aircraft engine plant, for example, an operation completion transaction will not be accepted unless critical process traceability data are included.

Since end users of a particular product will not know the details of which shop orders or materials were used to manufacture the product, another piece of information is required. This is usually a serial number, date of manufacture, lot number, or some other code that is tied to the product so that it can be traced back to a particular manufacturing order and its associated traceability information.

When orders close on the requirements file and the operations detail file, they are typically purged from the files after all other file updates (labor detail, cost detail, etc.) and reporting are completed. When traceability data is posted to these files, the required information, tied to the order, is passed to a history file as part of the order close and purging process.

Process Industry Considerations

Commonly recognized process industry groups include chemicals; rubber and plastic products; textile products; food and beverages; petroleum refining; stone, clay, and glass products; metal refining; paper products; lumber and wood products; and many others.

Frequently, practitioners from these industries cite the need for certain capabilities in their manufacturing data foundation because of the operating characteristics of these businesses. Some needs, such as the ability to handle by-product/co-product and reprocessing, which require specialized bill of material constructions and computer system logic, are discussed in the appendices. Others are considered here.

Note that some of the capabilities considered here are not necessarily unique to process industries. For example, the traceability requirements necessary in food and drug manufacture are equally as important in aircraft engine manufacture, hardly a process industry.

Frequently, the role of the bill of material versus that of the routing is less distinct in many process industries. The data required to define the product and process often combine the recipe/formula/foundation (bill of material) and the process sheet (routing). This is not dissimilar to Grandma's recipe for baking a chocolate cake: It lists both ingredients and baking instructions!

Routing Needs

Many process lines are capital intensive and equipment paced rather than labor paced. As a result, routings will contain data elements, such as line rates per hour, that are needed to define equipment constraints. Work center data elements needed for chemical reactors and vessels may describe an area like batch size parameters.

It is sometimes necessary to accommodate yield planning at individual routing steps. Yield losses through successive operations compound into a projected overall loss for a particular operation or manufacturing stage. An example of this can be found in specialty metals production, in which each step performed while fabricating ingot into rod results in some yield loss in pounds. (Of course, the metal is not really lost in large measure, since it's recycled through the melting furnaces.) The compound loss is measured by the pounds of rod netted from the pounds of ingot. At the operation level, it may be necessary to project the yield of rod expected based on projecting yield loss through the remaining operations. Semiconductor manufacturers report similar needs.

Product Sequencing and Campaigns

When the process lines and equipment are expensive and the clean-out/changeover/setup is sizable, it is common to sequence production in a natural progression from one product to another within a family. We might see this in the progression from lean to rich alloys of brass in a melt furnace before cleaning out and changing over to a different alloy family.

Similarly, the printing press used to print decorative sheets of walnut-grained plastic laminates would shift from wide to narrow widths and light to dark colors. As the print run progressed, the buildup of dirt and other impurities would be hidden by the progression. Once the walnut grain runs were complete, the press could be cleaned out and plates switched to the next pattern.

Data elements such as *product family code* may be required to support product sequencing needs. When potential costs and benefits are sizable, custom computer logic may be developed to sequence production. Otherwise, knowledgeable people will establish the schedule sequence.

Lot Traceability and Expiration

Many process industries, notably those involving food and beverages, drugs, cosmetics, and medical apparatus, are subject to government regulation, and must maintain records that detail the lot identification of materials used in the manufacture of these products. Many businesses follow this practice to protect themselves against liability.

For the most part, this business requirement does not affect the standard bill of material or routing for a product. However, it does make it necessary to maintain supporting inventory records that identify various lot numbers of an item on hand. In turn, any issues to manufacturing must provide the ability to maintain historical records of lots used. In general, the day-to-day operating systems that handle or generate transaction information must work with lot identification information.

Lot traceability systems provide capabilities similar to those found with standard bill of material processing systems. They make it possible to track the lots of material used in the manufacture of a particular item based on items such as lot number, serial number, or date/time of manufacture. This is similar to a bill of material explosion and can be both single level and indented, that is, identifying lots used at all levels of manufacture. Similarly, a lot *where-used* facility may also be available that permits tracking *up* to the items produced from a given lot of material.

Both approaches are frequently used in investigating the cause of an airplane accident. For instance, when cracks in an engine rotor were detected in a recent plane crash, the record-keeping systems allowed the engine manufacturer to identify the titanium ingot from which that particular rotor had been machined. Subsequently, other rotors produced from that same ingot were identified and tested.

Shelf life or lot expiration tracking systems also require supporting inventory record subsystems. Typically, they track lot creation dates and expiration dates and provide for first-in, first-out (FIFO) use of material as well as periodic aging reports used to predict material that is potentially expiring.

Potency or Concentration Considerations

The concentration of a material—for example, a 61 percent solution of hydrochloric acid (HCl) or the amount of active enzyme contained in a particular material—can be a data foundation consideration. Usually, the bill of material or formulation will specify the quantity required of an ingredient at standard concentration. A supporting subsystem or a knowledgeable person translating this information will be required to convert materials at other potencies or concentrations to standard equivalents.

In some instances, it may be possible and appropriate to process an ingredient prior to its use in the intended process to bring specifications to standard.

Variable Bills of Material

The actual bill of material used to produce an item may vary because of specifications or availability of ingredients or raw materials. Potency or concentration is certainly a factor. Cost or availability of virgin commodity materials, such as cathode copper, versus scrap material of less purity will affect the recipe for a furnace charge.

An extreme example can be seen in the very sophisticated custom software developed to optimize the melt mix for a specialty metal given the available raw materials and their specifications. To work, software of this type requires a large amount of attribute data about ingredient items.

Backflushing Considerations

Both process and repetitive industries rely on extensive *backflushing* for recording material or ingredient usage. Finished product or intermediate production reporting are used as appropriate to explode consumption.

Point-of-use information in the bill of material provides the needed linkage. Feeder stocks stored in holding tanks or materials stored at a particular point-of-use station on an assembly line are relieved based on reported production.

Discrete issues, receipts, and transfer transactions are routinely used to record movement in and out of controlled stores areas. Frequently, however, discrete transactions are inappropriate in flow shop production.

Making It Happen

The basic skeleton for the manufacturing data foundations is largely the same for both job shops and flow shops—bills of material, routings, related work center definition, and item master data. Within each, however, additional and/or different data elements may be appropriate to plan properly and control specific physical and operating characteristics.

Data foundations alone accomplish little. However, tied to the right planning, scheduling, and control software, they become a major element in a company's ability to succeed or fail in its business objectives. Whether the systems are simple or complex, the need for accurate and complete data foundations remains a major business requirement.

Making the planning, scheduling, and control systems happen is a subject greater than this book can properly address. It is important that those charged with creating and maintaining the company data foundations understand the specific data requirements necessary to support the planning, scheduling, and control systems. This includes the need to identify every data element that will be utilized by these systems and to assign the proper functional responsibility and accountability for maintenance control. Also required are the correct levels of data integrity and data file structure—topics covered elsewhere in this book.

Chapter 6
Modularizing the Bill of Material

At the outset, we said that it was important for a company to carefully structure its manufacturing data foundations to (1) support the manufacturing process and (2) support the type and variety of products a company wants to sell. In the previous chapters, we looked at the data foundation from the perspective of the manufacturing process. Now we turn to how a company's bills are structured to support the products it sells. This deals with the architecture of the bills, or how we choose to construct them. We must decide how to organize our underlying data base to support our forecasting and master production schedule needs. We may have a bill of material for each end item we produce, or we may arrange our bills to represent product families or models and the many options and features offered to meet a specific customer order configuration. Our choice makes a difference. In many cases, the latter choice is best handled by an architecture known as *modularized bills of material*.

Probably the best place to begin our explanation of modularization is to return to the Hunt House Restaurant and eat our way through the process.

The Hunt House Continental Restaurant decided to run a special Lobster Fest during the summer months, and it became a huge success. Each day, it featured six lobster variations. The lobster was baked and stuffed with crab, boiled and served with drawn butter, prepared in a thermidor sauce, added into their bouillabaisse, sautéed with lemon, butter, and herbs, or marinated in white wine and garlic and then mesquite grilled. Accompanying these entrées, customers had a choice of potato—baked, mashed, French fried, or the chef's renowned au gratin. They also could choose from a variety of vegetables—cauliflower mousse, braised watercress, gingered Chinese pea pods, carrots and pearl onions, green beans Napoli, or summer squash amandine.

In spite of the influx of people it expected to attract with the Lobster Fest, the Hunt House, as always, was particularly conscious of its unsurpassed reputation for excellent, prompt customer service. Harvey Allen was also concerned about keeping the restaurant's production costs as low as possible (as is everyone in business).

To meet these objectives during the Lobster Fest, Allen and his chef did some forecasting. They figured out, as best they could, what their customers probably would order. From experience, they decided that out of the 100 lobster dinners they expected to sell each day, they would receive orders for 20 boiled lobsters, 15 baked with crab, 15 thermidor, 20 mesquite grilled, 20 bouillabaisse, and 10 with lemon, butter, and herbs. They also figured that 35 people would order au gratin potatoes, 25 baked, 20 mashed, and 20 French fries. They thought that the vegetable options would be evenly divided among their customers. All plates, however, were garnished with the Hunt House's distinctive crab apple. What the Hunt House had in their limited restaurant environment was a product family—lobsters—with a good deal of raw materials involved, from herbs to vegetables, and a large number of end item configurations.

Of the first 50 customers that ordered lobster on day 1, 16 requested boiled, 20 mesquite grilled, 6 bouillabaisse, 4 baked with crab, 3 thermidor, and 1 lemon, butter, and herbs. Twenty-five ordered au gratin potatoes, 15 baked, and 10 wanted the fries. The vegetable choices were divided evenly among all but the carrots and pearl onions, which was chosen for only one dinner. Because these lobster and side dishes were all prepared in advance, Allen's customer service was 100 percent, but now Chef Milliken was getting a little concerned. The next two people who came in for the Lobster Fest both wanted the mesquite grilled. They heard from people leaving the restaurant that this was far and away the finest lobster they had ever tasted. The waiter had to explain to these customers that there was no more of the marinated lobster to grill. Allen's customer service dropped to zero. At the end of the day, as Harvey and Chef Milliken reviewed the receipts, they found that they had sold 20 boiled lobsters, 20 mesquite grilled, 14 baked, 15 bouillabaisse, 10 thermidor, and just 1 lemon, butter, and herbs. They also sold 35 au gratin potatoes (and could have sold more), 25 baked, 15 fries, and 5 mashed. The only vegetables left were 3 orders of green beans, 2 watercress, and 15 carrots and pearl onions.

Three days later, there were still 9 servings of lobster with lemon, butter, and herbs and 2 thermidors. Three customers arrived, having heard nothing but wonderful things about the Hunt House's Lobster Fest. All three thought that the lobster with lemon, butter, and herbs sounded "deelish." The waiter put in their order. At first, Chef Milliken was extremely pleased, but then he realized that the lobster with lemon, butter, and herbs was four days old. If he served it, those three people might become ill. He informed the waiter that in spite of the fact that the lobster with lemon and herbs hadn't sold, they had better not try to serve the remaining portions. He knew that even poor customer service was better than sick customers.

That night, Charles Milliken and Harvey Allen sat down to talk. By preparing the lobster and side dishes in advance, they were initially able to meet their objective of keeping their production costs low while providing excellent customer service. Unfortunately, a good portion of their inventory investment became obsolete. They realized that they hadn't approached this in the right manner.

They laid out all the recipes—all their bills of material for the different lobster dishes. Not too surprisingly, they discovered that lobster was common to all these recipes. They realized that it would be a lot easier to plan lobster needs in total, separate from the various ingredients that they combined with them when they configured-to-order any one of the end item recipes. The advantages were apparent. If they held their inventory at a semifinished goods level instead of at the end item level, they'd have a lower inventory investment, it would be easier to forecast lobsters in aggregate rather than forecasting all their possible configurations, and there would be less risk of obsolete inventory. The problem they faced was that configuring-to-order increased their lead times. It takes longer to prepare those different dishes when someone comes in and orders them. Chef Milliken's answer was to add a little extra cost by including a complimentary salad. He also thought that he could reduce some of the preparation time by making some of the sauces ahead of time. In making these adjustments, Chef Milliken planned to handle the increased lead times that his configure-to-order process would demand without jeopardizing customer service.

Like a lot of manufacturers, the Hunt House found that dropping down a level in its bill of material from its end item state made forecasting easier, lowered inventory investment, and actually increased perceived customer service.

Modular Bills of Material: Product and Business Considerations

Modular is a good thing for a product to be nowadays. As consumers, we see advertisements for modular stereo or home entertainment systems, modular telephones, modular homes, and modular computer systems. We've been trained to think of these modular products as being available in a wide variety of configurations. We configure our choice by selecting from a series of standardized units designed for use with one another. A modularized bill of material is a similar idea.

Let us now examine some of the issues that a company needs to consider in order to decide if modularization of the bill of material might apply to its environment. The process begins by understanding the type and variety of products sold. When there are a limited number of end items made from many intermediate items or components, as in Figure 6-1, the end item level is normally the forecast and planning level, and bills of material are maintained for each end item.

Another company may offer many end items made from a smaller number of intermediate items or component sets, as in Figure 6-2. This figure relates to our lobster example, with its many side dish options. Another example might be a particular model of automobile that has only a reasonable number of options to choose from but an endless array of highly featured and optioned end items. By dropping down to the basic option and feature level in the bill of material, the auto industry finds the natural forecasting and master schedule planning level. This and similar product examples are classic candidates for using the modularized bills of material discussed in this chapter.

As illustrated in Figure 6-3, another manufacturer might produce many items from a limited number of components. This is typical in many process manufacturing industries, such as those producing chemicals, metals, and textiles. In manufacturing T-shirts, for example, 10 or 12 different yarns may be used in manufacturing different styles of T-shirts. In the dyeing and finishing processes, 150 to 175 dyes and chemicals may be used. Even though there are fewer than 200 raw materials, when

Figure 6-1

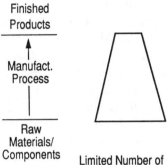

Finished Products

↑

Manufact. Process

Raw Materials/ Components

Limited Number of Items Made from Many Intermediate Items or Components

Figure 6-2

Many Items Made from a Smaller Number of Intermediate Items or Sets of Components

Figure 6-3

Many Items Made from Limited Number of Components

combined into all the various styles, sizes, and colors that a manufacturer must produce, a T-shirt manufacturer can easily make over 5,000 different end items. The results are similar to the auto/lobster examples. The end item level is not an effective forecasting and master schedule planning level. There are simply too many possibilities. Similarly, maintaining a unique bill of material for each end item becomes cumbersome.

Figures 6-2 and 6-3 characterize many manufacturers that offer large numbers of end item products. Their inability to forecast and plan each item effectively, as well as their desire to avoid zillions of bills of material, make them fertile ground for modularized bill concepts.

A connection often exists between the number of different end items sold and the delivery lead time offered to the customer. For example, a highly configured product may not be available off the shelf with zero delivery lead time. Very few manufacturers could afford to carry the inventory needed to operate in this manner.

This leads us to the second business issue that must be addressed in evaluating modularized bills of material. Where do we want to meet our customers? This refers to the marketplace impact of our delivery lead times to the customer. In the restaurant, Chef Milliken addressed a lengthening delivery time with a complimentary salad. Most industries, however, aren't fortunate enough to have such an easy solution.

Looking at the manufacturing process for the product represented in Figure 6-4, we see that its total cumulative lead time, from point of conception to completion date, is approximately 48 weeks. If, however, we had to go through this 48-week process every time we received an order for this product, it wouldn't take very long before our customers would be taking their business elsewhere, especially if our competition delivered comparable products in less time.

**Figure 6-4
Delivery Versus Manufacturing Lead Time**

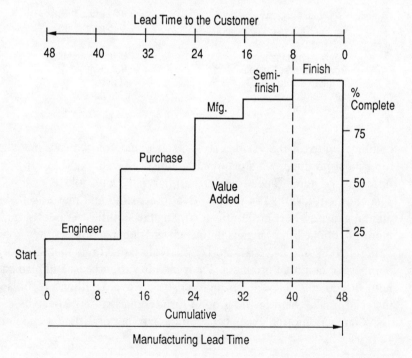

Modularizing the Bill of Material 105

For example, if an 8-week delivery to customer was competitive in our industry, we would need a forecast and master scheduling process that could pull product 40 weeks through manufacturing to a stage of semifinished goods. From this stage, we could complete the product in 8 weeks. In other words, our customer delivery lead times would be less than our overall or cumulative manufacturing lead times. This, in turn, would dictate certain things about the way we would design the architecture for our bills of material. We could orient bills toward lower-level items, such as longer lead time components, raw materials, intermediates, or subassemblies, from which we produce end items. Again, in this case, it's worth studying modularized bills of material.

In addition to considerations and trade-offs impacting forecasting, master scheduling effectiveness, and delivery lead times to customers, there are several other areas of concern. Using a hypothetical example similar to the model in Figure 6-3, let's look at the issues.

In Figure 6-5, we see the last three stages in a manufacturing process with optional choices. At the bottom level, customers choose one of five available items on one side and one of ten matching items on the other. This gives 50 different permutations of intermediate #2. We combine that in the next stage with the customer's choices of one of three items and one of four items. This makes possible 600 different configurations of intermediate #1. At the final level, the customer has two more decisions to make, choosing from one of two items and one of three. This makes a possible end item total of 3,600 different configurations. For this example, we can assume that the lead times for these manufacturing stages are two weeks for the bottom level, one week for the middle, and one week for the top.

If we wanted to stock end items to forecast, so that when a customer wanted one, it could be packed and trundled into his or her waiting van, our lead time to our customer would be zero. If, however, we wanted to stock second-level items to forecast and then make end items and first-level intermediates to customer order, delivery lead time to the customer would be two weeks. We could also choose to make both intermediate levels and end items to order, which would make delivery lead time to the customer four weeks.

Finally, we might decide to find a happy medium and create a hybrid strategy that could combine these different approaches. We might stock some of our most popular configurations at the end item level, available off the shelf, and some at lower, make-to-order levels. There are many choices.

**Figure 6-5
Alternative Strategies**

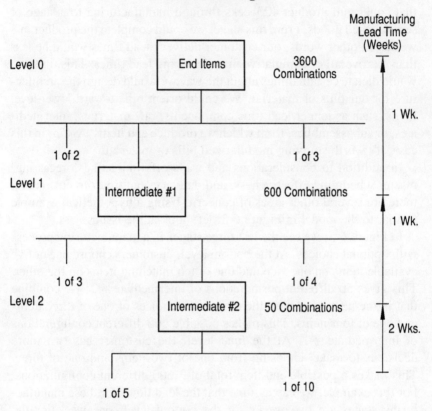

Modularizing the Bill of Material 107

How can we then determine which is the best choice for our company? What we want to examine is how each of these alternatives ranks according to important trade-offs we must consider. Our bill of material architecture is largely determined by our conclusions.

Any business strives to minimize delivery lead times to its customers. In our example, the best choice would be to stock end items. The worst delivery lead time would occur in alternative 3—making levels 0, 1, and 2 to order. Alternative 2 is the middle ground. These rankings are summarized in Figure 6-6.

Figure 6-6
Business Trade-offs

Bill of Material Structure
Must Consider Trade-offs...

Rank: B = Best W = Worst M = Middle	Alternatives		
	1	2	3
Customer Delivery Lead Time	B	M	W
Forecasting & Master Scheduling	W	M	B
Inventory Investment	W	M	B
* Shop/Work Orders Required	B	M	W
* Mfg. Costs	B	M	W
* Transactions Required	B	M	W
Ease of Order Entry	B	M	W
BOM's to Maintain	W	M	B

* Assuming lot-sized batch manufacturing.
  JIT environment would minimize differences.

Alternative Strategies	Delivery Lead Time to the Customer
1. Stock end items to forecast.	0
2. Stock intermediate #2 to forecast; make intermediate #1 and end item to order.	2
3. Make both intermediates and end item to order.	4

Another consideration is how our strategy would impact the forecasting and master production scheduling process. In this case, the worst alternative would be to stock end items to forecast. This would mean having 3,600 different items to forecast and plan at all times. The second alternative of stocking level 2 intermediates to forecast and making end items and level 1 intermediates to order would be a simpler choice, with potentially only 50 items to forecast. Making end items and both intermediates to order would be an even easier choice, since we wouldn't have to forecast any level 0, 1, or 2 combinations of components. It must be recognized, however, that in the latter two alternatives, we would have to forecast and plan any component items with lead times that exceeded our finishing or final assembly lead time.

From an inventory investment standpoint, carrying 3,600 different end items would be the most excessive choice. Making everything to order would be the least costly inventory approach.

In a lot-sized, batch, non-JIT environment, stocking all end items would probably lead to the least amount of factory work orders and associated paperwork, since we would make our products in economic order quantities rather than individually to order. In terms of manufacturing costs, as well as the number of transactions required in the factory, the same reasoning could be used (again in a non-JIT environment), with economic order quantities winning out over a make-to-order situation. In a make-to-order, work-order-controlled plant, we would need many more transactions to track production activity because there would be more work orders for smaller quantities. In JIT/TQC environments that are attacking areas such as paperwork and lot sizing, these comparisons would probably not be true.

The order entry process would also have trade-offs. It would be at its simplest with stocked end items. Each would have an item or catalog number, and customers would order from that number. Order entry probably would be more complicated under alternatives 2 and 3. It could be a *menu-driven* process, where a customer would make several selections, as in choosing options on a new car.

In terms of the size of our bill of material data base, it would be at its maximum if we permanently maintained individual bills for each of the 3,600 end items. Our data base would be much smaller if we kept only lower-level stocked items on file permanently, while constructing temporary end item bills when we had a customer order on hand. More on this later.

Figure 6-6 illustrates the trade-offs that must be considered when

Modularizing the Bill of Material 109

companies produce product families with large numbers of end items. It may be advisable to modularize the bills in instances where end item forecasting and planning are not practical, yet customers demand delivery in less time than cumulative manufacturing lead time.

MODULARIZING THE CCS PRODUCTS

Circumstances prompted the CCS Carpet Cleaner Company to consider modularizing their bills of materials. Business has really picked up since they decided to expand their product line. They recently introduced a heavy-duty model and a commercial model, both highly featured, optioned products. They have also discovered a global market for their carpet cleaners and have begun exporting their products. They now offer regular-duty carpet cleaners (for domestic sales only), heavy-duty carpet cleaners, and commercial carpet cleaners, some of which they export.

With the expanded product line, customers select specific cleaners from a number of predesigned options. There are five choices of capacity, four horsepower ratings, and two brush arrangements. Customers also choose domestic or export versions. If all the choices were available in any combination, there would be 80 end item configurations possible. (For purposes of simplicity, we have greatly limited the number of choices. Figure 6-7 shows the specific choices as described in the CCS catalog. Only 25 end item configurations are actually available.)

Figure 6-7
Carpet Cleaner Option Availability

	DOM.	EXPORT	HORSEPOWER				CAPACITY					BRUSHES	
			1.0	1.5	2.0	3.0	3P	4P	5P	5S	6S	3	5
0123 Standard	S	X	S	X	X	X	S	X	X	X	X	S	X
0124 Heavy Duty	O	O	X	O	O	X	X	O	O	X	X	S	X
0125 Commercial	O	O	X	X	O	O	X	X	X	O	O	O	O

S = Standard
X = Not Available
O = Option

P = Painted Tank
S = Stainless

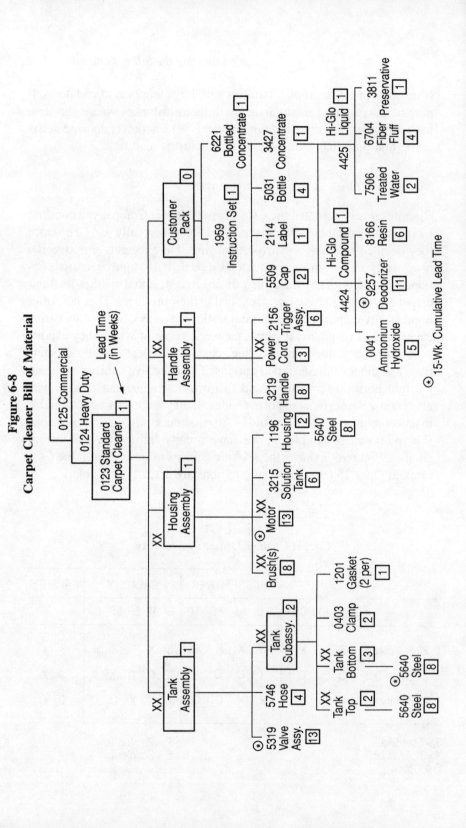

**Figure 6-8
Carpet Cleaner Bill of Material**

Now let's look at the expanded CCS Carpet Cleaner family bill of material. In Figure 6-8, we see the three different cleaners—the standard #0123, the heavy-duty #0124, and the commercial #0125. Each model is assembled from four first-level components; the variations show up in the second and third levels in the power cord choices, tank choices, motors, and brush specifications. We have placed XX on those optioned or featured parts, and we have included the level-by-level lead times. It is now possible to find the cumulative lead time for our carpet cleaner family. This is done by adding the different paths, or legs, of the bill. In this case, the longest paths are through the tank assembly/valve assembly leg, the tank assembly/tank subassembly/tank bottom leg, the housing assembly/motor leg, and the customer pack/bottled concentrate/concentrate/Hi-Glo compound/deodorizer leg, which all take 15 weeks to purchase and produce.

This is much easier to see when we tip the bill on its side and display lead times in parallel, as in Figure 6-9. This view clearly shows that if the factory shelves were bare, it would take 15 weeks for CCS to be able to deliver a customer order. It also means that should CCS want to deliver from stock, they would need a planning process that extended at least four months into the future in order to forecast marketplace requirements (longer to provide visibility on longer lead time items). The reason is simple: They must buy material today so that they can produce cleaners in week 16.

CCS offers their customers a 2-week delivery lead time, which compares favorably with competitive lead times in the industry. This means, of course, that they must have an on-hand or in-process inventory of components and a final assembly process that delivers in 2 weeks, much less time than their 15-week cumulative lead time.

For that to occur, CCS had to devise a planning and forecasting process using their master schedule to pull everything through manufacturing up to that 2-week point. To emphasize this, we've drawn a segmented line on Figure 6-9. For CCS to ship finished product in 2 weeks, everything on or to the left of the segmented line must be in stock or in process prior to the receipt of the customer order. Although there are 14 arrows touching the 2-week delivery line, in reality, there are more than 14 items that must be stock or in process, since some of these are optional items—for example, the domestic power cord (part #7114) and the export version (part #8321). There are also 5 tank subassembly part numbers, depending on the tank choices.

**Figure 6-9
Lead Times in Parallel**

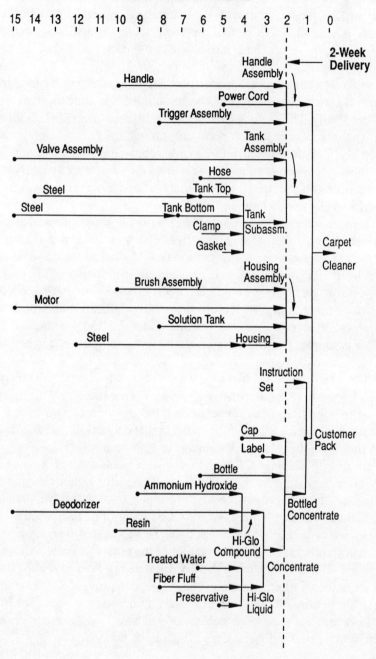

The CCS bill of material team then set up a series of bills that looked like those in Figure 6-10. The team knew that it made no sense to have a planning process that independently planned common components such as handles, trigger assemblies, valve assemblies, or hoses. To make any of their cleaners, they needed a matched set of all those components. Recognizing this, they created a new item called a *common parts* group and identified it as #CO99. Then they created a bill of material that allowed them to plan common components for their various cleaners as a matched set of parts in a coordinated fashion.

Figure 6-10
Modularized Bill of Material

* C099	COMMON PARTS	MOTOR OPTIONS	BRUSH OPTIONS
• 3219	Handle	* M010 1.0-Hp. Motor Option	* B003 3 Brush
• 2156	Trigger Assembly	• 6111 1.0-Hp. Motor	• 4315 Brush Assembly (3)
• 5319	Valve Assembly		
• 5746	Hose	* M015 1.5-Hp. Motor Option	* B005 5 Brush
• 3215	Solution Tank	• 1.5-Hp. Motor	• 4315 Brush Assembly (5)
• 1196	Housing		
• 1115	Customer Pack	* M020 2.0-Hp. Motor Option	
		• 5462 " " "	
		* M030 " " " Option	
		• 7703 " " "	

DOMESTIC/EXPORT OPTION		CAPACITY OPTIONS					
* D001	Domestic Option	* T003	3-Gal. Painted Tank Option				
• 7114	Power Cord	• 1910	"	"	"	"	Subassembly
* X002	Export Option	* T004	4	"	"	"	Option
• 8321	Power Cord	• 4721	"	"	"	"	Subassembly
		* T005	5	"	"	"	Option
		• 7350	"	"	"	"	Subassembly
* Pseudo Item Types		* T505	5	"	Stainless	"	Option
		• 9446	"	"	"	"	Subassembly
		* T506	6	"	"	"	Option
		• 8536	"	"	"	"	Subassembly

Item #CO99 is coded as a *pseudo* item on the item master file, as discussed in Chapter 2. A pseudo is the parent of an artificial collection of related components (common parts) that are grouped together for planning purposes. It is not an item that actually can be manufactured or inventoried. If we were to go into the CCS stockroom and look for #CO99, there would be no such item or assembly. It is an item number that exists for planning purposes only. When material requirements planning is run, it "blows through" the pseudo to the next stocking level in the bill—in this case, the actual components that must be issued to the factory.

Besides the common parts pseudo, CCS also created other bills for planning their options. For the power cord option, they would have two choices, domestic or export. For the tank or capacity option, there would be five choices. The motor option would have four choices and the brush option two. These are the options from which the expanded products can be configured. This is known as a *modularized bill of material construction*. What that means is that the bill has been broken or disentangled into separate predesigned building blocks, which can then be pieced together to create the finished products. By grouping these parts in this fashion, CCS can pick from them to configure the product that their customers want. In so doing, they have also been able to reduce the planning level from 80 possible (theoretical) configurations to 14 different items. By taking this step, they have attempted to structure their bills to support the product they offer and sell.

Several of the items created as pseudos actually contained only one component item. When asked why this was necessary, the bill of material team explained that they would cover that later, after we talked about integrating the bills and routings. So, more on that later.

This concept of modularization is commonly used where large numbers of end items are found. As mentioned previously, automobiles are a frequently cited example. When we order a car with power steering, the option may read as a power steering group or option. That item is more than likely a pseudo parent for a long list of component parts that are required to assemble a power assist for a car. Included would be all the components for the hydraulics, pumps, fittings, belts, and so on. Such bills of material are simply groups of items required in the end item when a particular option is specified. The term *option sensitivity* is used to describe this relationship. Modularization is the process of organizing components into bills of material based on their option sensitivity.

PLANNING BILLS OF MATERIAL

In a company that uses a modularized architecture to construct its bills, the *planning bill of material* usually becomes an important tool of the planning, sales, and marketing operations in top-level planning. The planning bill (also known as a *super bill*) provides a usage or forecast relationship between a product family or model and its options or features. This relationship is used to coordinate both the forecasting and master scheduling of these options.

In terms of CCS's #012X carpet cleaner family, made up of their standard, heavy-duty, and commercial cleaners, a planning bill would look like the one in Figure 6-11. Each available option is structured as a component to the #012X family or model item with a quantity per parent that represents its forecasted popularity or option percentage. Option percentages were developed as follows: The common components group is structured into a planning bill of material with a quantity per of 1 representing 100 percent usage. CCS projects that 50 percent of their market demand will be for their standard model #0123, 35 percent

Figure 6-11
Planning Bill of Material

012X Family

Master Scheduled Items		DOMESTIC/ EXPORT	MOTORS	TANK CAPACITY	BRUSH ARRANGEMENT
	100%	83%	50%	50%	91%
	Common Parts	Domestic	1.0 Hp.	3-Gal. Painted	3 Brush
		17%	10.5%	7%	9%
		Export	1.5 Hp.	4-Gal. Painted	5 Brush
			30.5%	28%	
			2.0 Hp.	5-Gal. Painted	
			9.0%	7.5%	
			3.0 Hp.	5-Gal. Stainless	
				7.5%	
				6-Gal. Stainless	

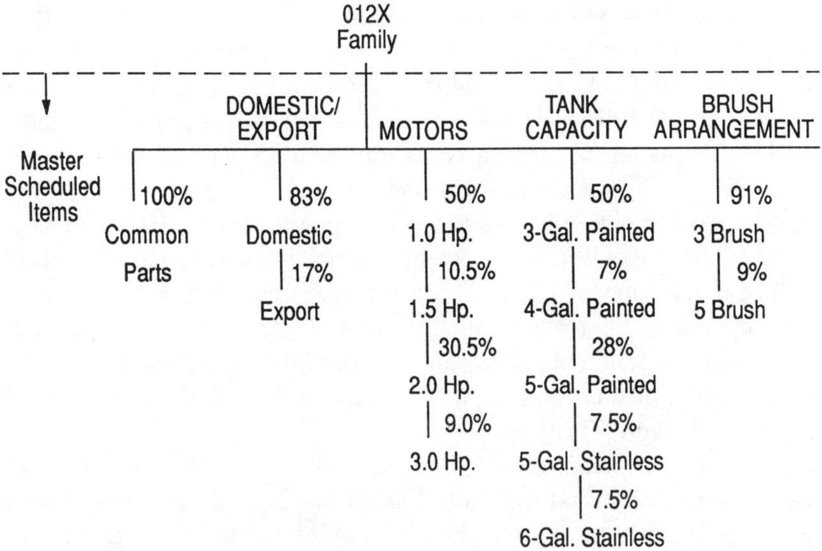

for model #0124, with model #0125 making up the final 15 percent (see Figure 6-12). These projections are made by their sales and marketing groups, based on both analysis of past sales history and knowledge of market trends.

CCS knows that all its standard models will be made for domestic sale. It has also been decided that of those #0124 heavy-duty cleaners they make, 60 percent will be the domestic variety and 40 percent will be for export. Of their #0125 models, 80 percent will be sold at home and 20 percent will be exported. They also forecast that of the two possible motor configurations for the #0124, 30 percent of their customers will choose the 1.5-horsepower motor and 70 percent will take the larger 2-horsepower motor. For the #0125, they forecast that 40 percent of their customers will want the 2-horsepower motor and 60 percent will want the 3-horsepower motor. The standard #0123 has only one choice—the 1-horsepower motor.

In terms of tank capacities, again, model #0123 has one choice, the 3-gallon painted tank. For model #0124, the forecast anticipates that 20 percent of sales will be for the 4-gallon painted tank and 80 percent will be for the 5-gallon painted tank. Model #0125 is divided 50-50 between the 5-gallon and 6-gallon stainless tanks. The optional brush arrangement is only available on the commercial cleaner, and CCS expects to sell 40 percent of the 3-brush model and 60 percent of the 5-brush model.

Figure 6-13 shows how these forecasted mix relationships are then used to calculate percentages for inclusion in the planning bill of material. For each 1,000 units sold over the next planning horizon, CCS expects to sell 500 model #0123, 350 model #0124, and 150 model #0125. Applying mix forecasts, we find that 83 percent of their business will be for domestic sales and 17 percent for export. Of their motors, 50 percent will be of the 1-horsepower variety, 10.5% percent will be 1.5 horsepower, 30.5 percent will be 2 horsepower, and 9 percent will be 3 horsepower. The tank options were figured to be 50 percent 3 gallon painted, 7 percent 4 gallon painted, 28 percent 5 gallon painted, 7.5 percent 5 gallon stainless, and 7.5 percent 6 gallon stainless. The brush option mix is forecast at 91 percent sold with 3 brushes and 9 percent sold with 5 brushes.

Of course, a planning bill of material that looks like Figure 6-11 would never end up on the shop floor. It has been devised purely as a planning and forecasting tool in the modularized bill of material construction we are illustrating.

Modularizing the Bill of Material   117

Figure 6-12
Market Demand Analysis

UNIT DEMAND		DOM.	EXPORT	HORSEPOWER				TANK CAPACITY					BRUSHES	
				1.0	1.5	2.0	3.0	3P	4P	5P	5S	6S	3	5
50%	0123 Stand.	100	0	100	–	–	–	100	–	–	–	–	100	–
35%	0124 H-D.	60	40	–	30	70	–	–	20	80	–	–	100	–
15%	0125 Comm.	80	20	–	–	40	60	–	–	–	50	50	40	60

P = Painted Tank
S = Stainless

Figure 6-13
Internal Planning Percentages

% DEMAND		TOTAL UNITS	DOM.	EXP.	HORSEPOWER				TANK CAPACITY					BRUSHES	
					1.0	1.5	2.0	3.0	3P	4P	5P	5S	6S	3	5
50	0123 Stand.	500	500	–	500	–	–	–	500	–	–	–	–	500	–
35	0124 H-D.	350	210	140	–	105	245	–	–	70	280	–	–	350	–
15	0125 Comm.	150	120	30	–	–	60	90	–	–	–	75	75	60	90
	Totals	830	830	170	500	105	305	90	500	70	280	75	75	910	90
	(%)		83	17	50	10.5	30.5	9	50	7	28	7.5	7.5	91	9

At CCS, planning with modularized bills of material has reduced the number of items to be forecast and planned from 80, the theoretical number of end items, to 14. That's the number of parts groups, components, and subassemblies visible in Figure 6-11 below the segmented line. Forecasting for these items should be more accurate than forecasts for 80 individual end items.

This reduction from 80 items to 14 items is pretty meager compared to many real-life product examples. A hydraulic lift truck manufacturer might report 1.8 million end items possible when selecting from 200 to 300 option items. The mathematics of permutations and combinations is working here! For many companies, modularized bills are the only practical way to reduce the number of items that must be planned to a reasonable number.

Planning bills are used in master scheduling to project mix needs and to drive rescheduling when the actual mix varies from the projected. As we will see later in this chapter when we discuss the order entry process, planning bills can also be used to detail and validate a customer's order configuration. We will also discuss *master* bills and their relationship to planning bills at that time.

As a forecasting and master scheduling aid, the usefulness of planning bills of material varies. Certain product situations are good fits; others less so. Planning bills of material make a great deal of sense when the family volume of a product is reasonable or large and the forecasted option percentages are also reasonable. In addition, they work well when the option mix is stable and the customer demand is smooth. They don't work well when the family volume is small—for instance, if a company makes only eight a year of a given family. They also don't work well when the option forecast is small—for example, 1.6 percent of the total—or where the option mix varies dramatically, or when the customer demand is typically uneven.

Other troublesome situations can also occur. Option percentage or popularity can vary over time, with seasonality a common cause. There may not be a ready solution other than to use the date effectivity techniques associated with engineering change, which we will discuss in Chapter 8. These may also be cumbersome.

Similarly, option percentages may vary by sales territory. If this condition is further complicated by seasonal considerations, planning bills may have to be developed by territory. In these situations, an amalgamated planning bill at the total family sales level may prove ineffective for both forecasting and planning.

The process used at the CCS Carpet Cleaner Company to modularize their bills of material began with the identification of their product family (the #0123 family) and the options that they wanted to offer their customers. The bill of material team determined the option sensitivity—what component items were needed when a customer specified a certain option. Then they created planning bills of material as an approach to forecasting and planning their options and features. Once that had been accomplished, they had to create a final assembly interface so that they could build the finished product to customer requirements.

By going through the analysis, they simplified their forecasting and planning and planned at a level in the bill of material that made sense for their operations. It was at this point that they determined that a modularized bill of material offered advantages in their environment.

Although we have illustrated these modularized concepts and planning bills of material with an assembled product—the carpet cleaner—many process industries also have potential for applying them. Packaging and labeling permutations increase the number of end items in pharmaceutical products. The same is true in the chemical industry. Where delivery lead times permit, holding inventory at the semifinished or bulk stage until the customer has specified the precise end item configuration enables a company to enhance customer service and reduce inventories.

One frequent benefit of applying the modular technique is that it forces a company to examine why it offers so many variants. One well-known pharmaceutical company reduced its packaging variants from 3,500 to 1,400 with no loss in sales.

INTEGRATING ROUTINGS AND THE BILL OF MATERIAL

Once CCS decided to modularize, they had to figure out how to use the bills of material and routing processes to support a modularized architecture. To do this, they returned to where they began. Look again at the parallel bill of material lead times in Figure 6-9. The segmented line drawn at the two-week competitive delivery mark had some implications not previously discussed. With their stocking level at two weeks, CCS was now able to make handle assemblies, tank assemblies, housing assemblies, and customer packs to customer order after the receipt of that order and still deliver within customer lead time. These items did not have to be in stock or in process before the company had a customer order in hand.

These subassemblies qualified as what were previously called *phantom* items in Chapter 2. Instead of stocking these subassemblies in advance, the company would now treat them as transient items, building them on the fly. As mentioned earlier, a phantom is an assembly or intermediate built and immediately consumed in the assembly or processing of its parent. Typically, the lot size of the phantom is discrete or lot for lot, and it carries a zero lead time on the item master file. In material requirements planning, this causes requirements to blow through phantom levels, so that requirements for its components equal needs for its higher-level parent.

By creating these phantoms, CCS has, in the parlance of JIT/TQC, *flattened* its bill of material, because it has functionally eliminated a level in the carpet cleaner bill. Actually, the company has flattened its manufacturing process by *flowing* product through the shop more directly, rather than processing in fits and spurts through intermediate stocking levels of manufacture. Reducing the number of bill of material levels becomes even more critical in JIT/TQC environments, where an unimpeded, swift flow is of great importance. These phantom items taught CCS how routings and bills of material can be integrated when *flat* bills of material exist.*

A caution should be mentioned here that certain industries need to keep in mind. It has to do with service or spare parts, intercompany supply items, direct sales of intermediate items, and the like. In CCS's flatter bill, the tank assembly does not exist as an item number or level in the bill of material. It may be necessary to supply that tank assembly item as a spare or other independent demand item. Such demands normally are serviced from stock. The dilemma that surfaces in these cases is software related. The item master file of many software packages allows us to put a single item-type code on an item. This code usually indicates whether a part is a purchased item, of manufactured item, a stocked item, a pseudo, or a phantom. These codes are usually mutually exclusive.

At CCS, the tank assembly illustrates the problem. The bill of material team has decided to treat the tank assembly as a phantom and make it as they go, in response to customer order. The spares people, however, want these same parts as a stocked item. The problem is that the company wants to treat one item number on the item master file as both a

---

* In this discussion, we will assume flat bills of material, where intermediate levels have been eliminated. Some companies choose to retain the levels in the bills of material but encode them as phantoms on the item master file. It is sometimes easier to do the latter because of drawing systems linkages, routing construction, and other existing company practices.

phantom and a stocked item, while the software sees this as an either/or situation. Actually, this is fairly common. On the one hand, we want to treat a part as a *blow-through* item; on the other hand, we want to stock it. In some cases, because of the constraints within the software itself, it becomes necessary to supply a unique item number for the item demanded from stock. For example, companies with significant spare parts demands may suffix the phantom item number with an s, indicating spares, and code this item number as a stock item. Other solutions may also be possible, but it is important to be aware of potential difficulties.

With that caveat aside, we can proceed to the relationship between a modularized and flat bill of material and the routings. Undoubtedly, an early question that arises is, How does a company actually manufacture anything in a modularized environment involving phantoms that don't appear in the bill and pseudo types that don't really represent anything that can actually be made? What we will illustrate now is how the CCS Carpet Cleaner Company has been able to assemble an integrated set of data that draws from both the bill of material and the routing to document the manufacturing process. To integrate this data, we'll look at bills of material coding, such as point of use, lead time offset, line sequence number, and the parent operation number linkages in the routings and bills of material.

Because of the large number of possible end items that would have to be carried on the item master file, CCS decided against maintaining part numbers at the end item level. Again, the CCS situation is a deliberately simplified example with only 80 possible end items. One might envision a more extreme situation, such as the aforementioned example of highly optioned lift trucks. The process of maintaining part numbers at the end item level when there are millions of possible end item configurations is mind-boggling, and few companies do it.

Nonetheless, when the decision to modularize was made at CCS, one question heard frequently in the hallways was, If part numbers at the end item level don't exist, what about the routings that describe how to make that item? Since routings are attached to part numbers, eliminating the end item part numbers meant that there couldn't be a routing at the end item level. CCS handled this by incorporating the equivalent data elements into the modularized bills and associated routings. The same potential routing problem exists with lower-level items, such as intermediates and subassemblies. As we make bills of material shallower, or

Figure 6-14
Routing/Process Information

(BEFORE) 0123 Carpet Cleaner Final Assembly Routing

OP.	DESCRIPTION	POINT OF USAGE
10	Assemble Handle and Housing Assemblies	FA-3
20	Attach Tank Assembly	FA-3
30	Connect Hose to Housing	FA-3
40	Insert Customer Pack	FA-3
	2927 Tank Assembly	
10	Attach Valve Assembly to Tank Subassembly	SA-5
20	Attach Hose	SA-5

(AFTER) C099 Common Parts

OP.	DESCRIPTION	POINT OF USAGE
10	Details of Handle and Housing Assemblies	—
80	Attach Valve Assembly to Tank Subassembly	SA-5
90	Attach Hose	SA-5
100	Assemble Handle and Housing Assemblies	FA-3
110	Attach Tank Assembly	FA-3
120	Connect Hose to Housing	FA-3
130	Insert Customer Pack	FA-3

flatter, by eliminating levels, we further disturb the existing routing or process instruction data base.

Figure 6-14 shows the before-and-after routing information in the CCS data base. The routings, formerly linked to the #0123 carpet cleaner and the #2927 tank assembly, have been combined and linked to the #C099 common parts item. The point-of-use information describes the subassembly and assembly areas where the manufacturing is performed.

It is possible to tie the routing to #C099 in this example because all carpet cleaners follow the same subassembly/assembly process. Some products, however, with particular configurations require *modularized* routings on the data base. For example, if the stainless steel tank on the commercial cleaner required an extra or different step in the process to attach a securing wire to the hose on the tank assembly, the routing would have to be modularized as well, because one or more routing operations would be specific to stainless steel tank configurations and not required for painted tank choices. An operation 95 would be attached to both #TS05 and #TS06 stainless tank options. A customer order for a stainless tank configurations would get a routing consisting of all routing steps tied to a #CO99 common parts item, plus the extra process step associated with either the #TS05 or #TS06 option.

Point-of-use data in the bill of material tells us the proper place materials are to be delivered in the factory (see Figure 6-15). By

Figure 6-15
Point of Use Information

To FA-3
↑
Phantom 3-Gallon
Tank Assembly
↑
C099 Common Parts Routing

OPERATION	DESCRIPTION	POINT OF USE
—		
—		
—		
80	Attach Valve Assembly to Tank Subassembly	SA-5
90	Attach Hose	SA-5

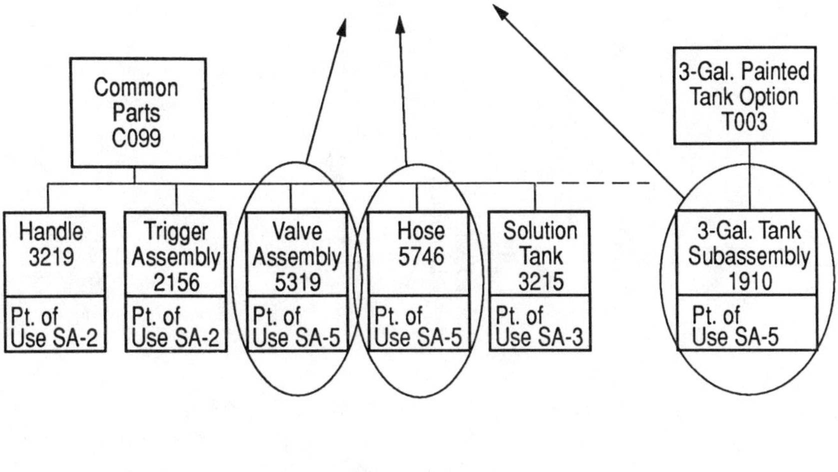

Common Parts C099

- Handle 3219 — Pt. of Use SA-2
- Trigger Assembly 2156 — Pt. of Use SA-2
- Valve Assembly 5319 — Pt. of Use SA-5
- Hose 5746 — Pt. of Use SA-5
- Solution Tank 3215 — Pt. of Use SA-3

3-Gal. Painted Tank Option T003
- 3-Gal. Tank Subassembly 1910 — Pt. of Use SA-5

Figure 6-16
Lead Time Offset Information

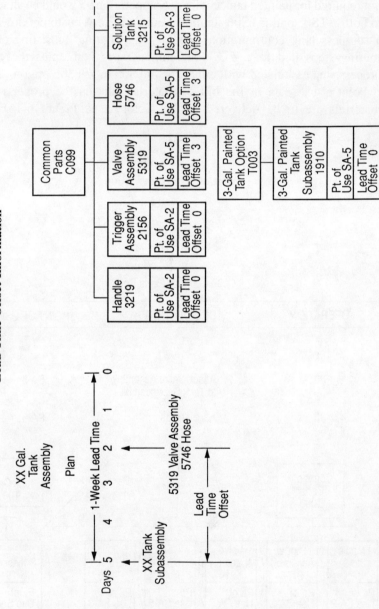

including the point of use on the bill, CCS makes sure that items in the common parts, as well as option-sensitive items, find their way to the correct work center. Both the #5319 valve assembly and the #5746 hose from the common parts group are used at SA-5, along with the #1910 3-gallon tank subassembly, to produce the phantom 3-gallon tank assembly.

If it were necessary to modify the time of need for an item, the *lead time offset* could also be included on the bill. This element can be used to control when materials are issued to a work center as well as determining the need date when planning, purchasing, or manufacturing a particular component (see Figure 6-16). Let us assume, for example, that the #5319 valve assembly and the #5746 hose were not needed at the SA-5 tank assembly area until the preparation had been completed on the matching tank assembly. The lead time offset could be used to control this timing.

Lead time offsets or their equivalent have long been used by manufacturers of major equipment, where there are long assembly sequences. They are also frequently in use—often expressed very precisely in hours—in flow environments such as repetitive assembly, where the whole process occurs within hours.

Sometimes we need to structure a given component into a parent more than once because it is used at different points of use, at different times in the process, or in different parent operations. Using the #1910 tank subassembly as an example, notice the coding assigned the #1201 gasket in Figure 6-17. Because of the different points of use and lead time offsets involved in its two structures, it cannot be structured just once into the #1910 parent. The line sequence number is part of the unique *key* in the product structure record. At CCS, the software will not allow two line sequence *010's* in a given parent, but a given component can be structured into a parent any number of times if unique line sequence numbers are assigned.

Line sequence numbers are sometimes used to sequence components in the desired order on a bill of material, frequently in order of use. They are also used on occasion to refer to the balloon or find number on drawings.

Parent operation number data in the bill of material are often used to integrate the bill of material and the routing. This allows CCS to specify the routing operation at which a particular component is used. Figure 6-18 shows this relationship for the components required to manufacture

Figure 6-17
Line Sequence Information

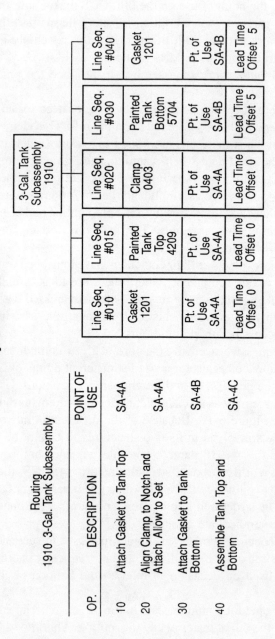

Routing
1910 3-Gal. Tank Subassembly

OP.	DESCRIPTION	POINT OF USE
10	Attach Gasket to Tank Top	SA-4A
20	Align Clamp to Notch and Attach. Allow to Set	SA-4A
30	Attach Gasket to Tank Bottom	SA-4B
40	Assemble Tank Top and Bottom	SA-4C

Figure 6-18
Parent Operation Number Linkage

Routing

C099 Common Parts

OP.	DESCRIPTION	POINT OF USE
–		
–		
–		
80	Attach Valve Assembly to Tank Subassembly	SA-5
90	Attach Hose	SA-5

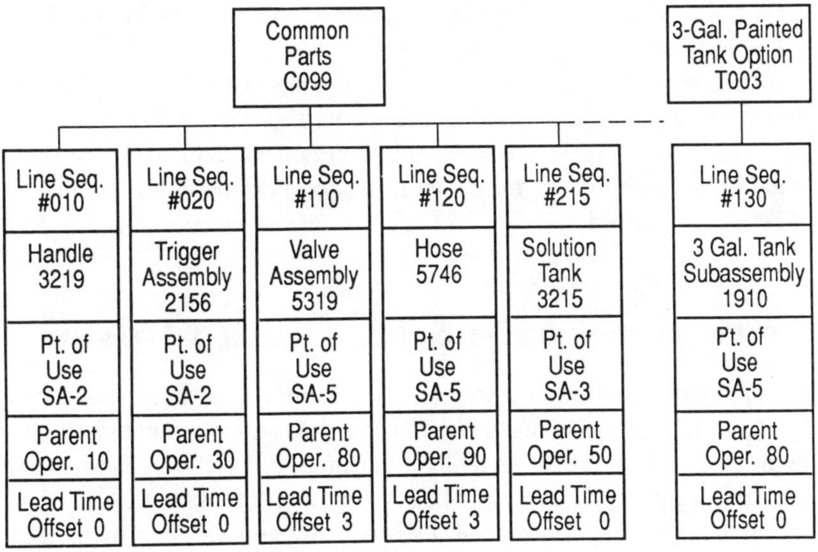

		Common Parts C099				3-Gal. Painted Tank Option T003
Line Seq. #010	Line Seq. #020	Line Seq. #110	Line Seq. #120	Line Seq. #215		Line Seq. #130
Handle 3219	Trigger Assembly 2156	Valve Assembly 5319	Hose 5746	Solution Tank 3215		3 Gal. Tank Subassembly 1910
Pt. of Use SA-2	Pt. of Use SA-2	Pt. of Use SA-5	Pt. of Use SA-5	Pt. of Use SA-3		Pt. of Use SA-5
Parent Oper. 10	Parent Oper. 30	Parent Oper. 80	Parent Oper. 90	Parent Oper. 50		Parent Oper. 80
Lead Time Offset 0	Lead Time Offset 0	Lead Time Offset 3	Lead Time Offset 3	Lead Time Offset 0		Lead Time Offset 0

the phantom tank assembly, which is the result of performing operations 80 and 90 of the #C099 common parts routing.

Parent operation number linkages are also helpful when *backflushing* techniques are used to relieve component inventories. If the elapsed time in manufacturing is substantial, it's difficult to wait until production is complete to report component usage. The delay makes reconciling component inventories virtually impossible. Provided the parent operation number linkage is maintained in the bill of material, component inventories can be updated as production is reported operation by operation.

This linkage is also useful in evaluating work-in-process inventory. If it can be determined that operation 80 has been completed on a particular customer order, both standard labor and material costs to that in-process unit, can be assigned.

The #T003 3-gallon painted tank option seen in Figure 6-18 allows us to revisit a question posed earlier in this chapter. We wondered why it was necessary to create a pseudo item (#T003) with only one component, in this case the #1910 3-gallon tank subassembly.

That can happen when the same component item is used in more than one family of products and becomes structured in more than one master-scheduled option. The pseudo item bill of material record permits us to carry such things as different points of use, parent operation numbers, and lead time offsets on the same item.

Conversely, if the component item is unique to a product family and these various bill of material data elements aren't needed to cost, plan, or control the finishing process, the component item itself can be a stand-alone master-scheduled item. No pseudo item level, such as #T003, is required.

THE FINISHING SCHEDULE OR FINAL ASSEMBLY INTERFACE

At CCS, with their modularized bills and routings, the final assembly or finishing schedule is actually used to schedule and manufacture the end item. Once a firm order, normally a customer order, has been received, the order entry process complies the proper options, features, and attachments needed to produce the finished item requested by the customer (see Figure 6-19). In a discrete unit production situation, the scheduler selects the orders for manufacture, produces the appropriate level of manufacturing documentation, and releases the component pick lists, which are summarized by component and sequenced by stores location. The parts are then delivered to the point of use and assembly finally commences.

In many factories, however, little documentation is required on the factory floor beyond skeletal details of what product to produce. The more repetitive the process, the less likely that detailed process information will be required. Pick lists with discrete issues and associated

**Figure 6-19
Finishing/Final Assembly Interface**

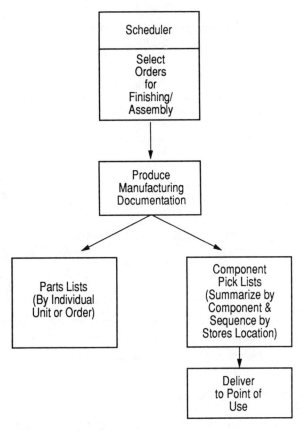

paperwork are not used. Components are transferred to point-of-use locations and are relieved with backflush techniques.

CCS produces a fairly complete set of factory documentation for assembly personnel. Detailed parts or pick lists and routing instructions are provided for each order (see Figure 6-20). A reference drawing, #R4260, has also been noted in routing steps 80 and 90 and is provided to the assemblers in area SA-5. CCS continues to find it useful to communicate assembly views, tolerance, and other information. It is important, in this instance, to have a reference drawing for a phantom assembly, the tank assembly, which has no item number on the item master file and which does not exist as a level in the bill of material. The

Figure 6-20
Factory Documentation

Today: Week 10

SA-5 Subassembly   Heavy-duty Export, Req'd. Date: Week 12
Cust.: 1234   10   2.0-Hp. Motor, 5-Gal. Tank
Assembly Parts List

REQ'D.	ITEM	DESCRIPTION	PARENT OP.
10	5319	Valve Assembly	80
10	5746	Hose	90
10	7350	5-Gal. Tank Subassembly	80

Assembly Routing

OP.		COMPONENTS		REFERENCE DRAWING
80	Attach Valve Assembly to Tank Subassembly	5319	Valve Assembly	R4260
		7350	5-Gal. Tank Subassembly	
90	Attach Hose	5746	Hose	R4260

#R4260 drawing actually refers to the routing steps, which, when completed, will result in the phantom item. At CCS, the drawing system has been *uncoupled* from the bills of material on the computer.

ORDER ENTRY CONSIDERATIONS

The planning (or *super*) bill of material usually found in a modularized bill of material environment can also be an effective tool in customer order entry. Some companies develop a *menulike* order entry selection process, using the options and feature items of the planning bill. Figure 6-21 illustrates the process, following customer order #1234 for ten heavy-duty cleaners with a ship date of week 12. Since the order is for an #012X carpet cleaner, this order would automatically trigger a demand for ten sets of common parts. The order entry process would then take the customer through a series of lien item choices of features and options. In this case, all the cleaners were of export variety with the 2-horsepower engines and 5-gallon painted tanks.

Companies with high-volume order entry situations can automate with custom order entry systems. Logic may be incorporated to prevent

Modularizing the Bill of Material 131

invalid combinations of options. For example, a customer cannot order a heavy-duty towing option on an automobile with a small displacement engine.

CCS might also approach order entry by assigning a special catalog number to each of the possible cleaner configurations. For example, the CCS catalog (see Figure 6-7) could list the #0124 heavy-duty cleaners as model #124-XL4, #124-XH4, #124-DL4, #124-DH4, #124-XL5, or #124-XH5. The #124 would refer to the heavy-duty model, the X or D would identify the export or domestic model, the L or H would refer to the high- or low-powered engine, and the 4 or 5 would identify the size of the tank. However, if CCS were to use a *smart* catalog number approach, their software and order entry mechanism would be designed to translate that catalog number into the internal item

Figure 6-21
012X Carpet Cleaner Order Entry

Customer: _____   Order No.: 1234   Req'd. Wk. 12
                                                                                           Date:

Order Detail

LINE ITEM	ITEM NO.	DESCRIPTION	QTY.	STANDARD
001	C099	Common Parts	10	ALL
Pick One				
002	D001	Domestic	___	0123
002	X002	Export	10	
Pick One				
003	M010	1.0-Hp. Motor	___	0123
003	M015	1.5- " "		
003	M020	2.0- " "	10	
003	M030	3.0- " "		
Pick One				
004	T003	3-Gal. Painted Tank Option	___	0123
004	T004	4- " " " "	___	
004	T005	5- " " " "	10	
004	T505	5-Gal. Stainless " "	___	
004	T506	6-Gal. " " "	___	
Pick One				
005	B003	3-Brush Option	10	0123,4
005	B005	5-Brush Option	___	

numbers that make up specific configurations. Systems of this type are commonly referred to as *configuration* systems.

Suppose that a customer chose the #124-XH5. Once the order was entered, the customer's configuration choices would be automatically translated into customer order details used internally. In Figure 6-22, line item 1 is #C099 common parts. Since #C099 is a pseudo item, the CCS system would blow through the #C099 bill of material, pick up its components and their point of use, parent operation, and lead time offset, and apply that information to the required dates on the order. In essence, we would create an order bill of material for this order that simply said, "This is exactly what our customer wants."

It's possible that this particular configuration would never be used again. This reinforces the idea that with many possible end item configurations, it is not only impractical, but unnecessary, to have all the

Figure 6-22
Customer Order Detail

Customer Order
Order No.: 1234   QTY.: 10   Request: 13   Ship: 12   Start: 10
Description 124-XH5
Heavy-Duty, Export, 2.0-Hp. 5-Gal. Painted Tank

LINE ITEM	ITEM QTY.	ITEM NO.	DESCRIPTION	ITEM TYPE	POINT OF USE	PARENT OPERATION	LEAD TIME OFFSET
001-00	10	C099	Common Parts	Pseudo	—	—	—
001-01	10	3219	Handle	Stock	SA-2	10	
001-02	10	2156	Trigger Assembly	Stock	SA-2	20	
001-03	10	5319	Valve Assembly	Stock	SA-5	05	3
001-04	10	5746	Hose	Stock	SA-5	07	3
001-05	10	3215	Solution Tank	Stock	SA-3	50	
001-06	10	1196	Housing	Stock	SA-3	60	
001-07	10	1115	Custom Pack	Pseudo	—	—	—
001-08	10	1959	Instruction Set	Stock	FA-3	130	4
001-09	10	6221	Bottled Concentrate	Stock	FA-3	130	4
002-00	10	X002	Export Option	Pseudo	—	—	—
002-01	10	8321	Power Cord	Stock	SA-2	20	
003-00	10	M020	2.0-Hp. Motor Option	Pseudo	—	—	—
003-01	10	5462	2.0-Hp. Motor	Stock	SA-3	60	
004-00	10	T005	5-Gal. Painted Tank Option	Pseudo	—	—	—
004-01	10	7350	5-Gal. Tank Subassembly	Stock	SA-5	80	
005-00	10	B003	3-Brush Option	Pseudo	—	—	—
005-01	30	4315	Brush Assembly	Stock	SA-3	40	

different possible end item bills of material on file. An order bill of material usually exists for the life of the order. Once it's completed, the bill of material is normally archived for historical record purposes, warranties, and so forth.

To assist in the order entry process, menu selection, and configuration control, some companies create and maintain a *master* bill of material. The master bill may contain product options that are either unpopular or custom engineered to order. Although valid options for the product, these items are not forecast and thus would not appear on the planning bill at all. If they did, they would have 0 percent forecast. In effect, the planning bill of material is a subset of the master bill of material.

The need for both a master bill and a planning bill may be alleviated when software provides both a quantity per and a forecast percentage in the planning bill of material parent/component record. When available, a given option can have a quantity per of 1 (indicating that one is required when this option is selected) even though the forecast is 0 percent.

ADD/DELETE BILLS OF MATERIAL

A technique often called the *add/delete bill of material* is occasionally used. In these add/delete environments, a variation on a standard model is developed, rather than repeating the entire product structure, by simply deleting unneeded component items and adding the replacements (see Figure 6-23). A shorthand way to describe a #0124DL4 is to create a bill of material that has as one of its components the standard #0123, except that the 3-gallon tank assembly and 1-horsepower motor are deleted and a new tank assembly and motor are added.

However, we recommend the use of modularized bills because of the way current time-phased planning systems work. Modularized bills are *add only*; there are no deletions.

If the system were to explode through the bill of material and find deletes, it is possible, because of timing and quantity differences, that negative requirements might exceed the positive requirements for an item and create a nonsense situation. Further, using the add/delete mechanism creates complex bills of material with a potential for inaccuracy. In addition, the master schedule planning approach requires end item scheduling and becomes tough to manage properly. In today's system environments, add/delete bills are not recommended.

Figure 6-23
Add/Delete Bill of Material
Carpet Cleaner Comparison

MODEL	DOM./ EXP.	MOTOR	TANK	BRUSH OPTION
0123	DOM.	1.0 Hp.	3-Gal. Painted	3
0124DL4	DOM.	1.5 Hp.	4-Gal. "	3

Components

PARENT	ITEM #	QTY. PER	DESCRIPTION
0124DL4	—		Carpet Cleaner
	• 0123	+1	Carpet Cleaner
	• 2927	−1	3-Gal. Tank Assembly
	• 1527	+1	4-Gal. " "
	• 6111	−1	1.0-Hp. Motor
	• 2740	+1	1.5- " "

MODULARIZATION IN PERSPECTIVE

It should be obvious by now that modularized bills of material can be an attractive approach in the right environment. They can be an aid in forecasting and master scheduling and can reduce the amount of maintenance of the bills of material and routings themselves. It must be understood, however, that their use, as opposed to end-item-oriented bills, has potentially widespread implications throughout a company. Some caution must be taken when a company is deciding whether or not to modularize its bills of material.

Many companies readily grasp the idea of modularizing into common versus unique materials and understand how planning bills of material can prove helpful—issues at the tip of the modularization iceberg. However, only about one-tenth of an iceberg's mass is above the surface. Not easily seen are the other factors involved—those that lie below the surface (see Figure 6-24).

If a company is considering modularizing its bills, it is essential that it has a firm handle on the technical capabilities necessary in its system to incorporate areas such as point of use, lead time offset, line sequence numbering, and parent operation linkages or their equivalents. These are particularly important in flat bill environments, as well as when phantoms and pseudos are involved.

A company must also be able to support the modularized architecture through various application considerations. Modularized bills will invariably impact sales literature, the order entry process, and how one promises delivery to the customer. They will also affect product costing, pricing, and invoicing, especially when a company chooses to eliminate levels in the bill. It's hard to roll up a standard cost for an end item if there is no item number or bill of material for it. These company processes are typically reoriented to cost, price, and invoice products by the options selected for the finished configuration, as in the example of the automobile.

**Figure 6-24
Modular Bill of Material Implications**

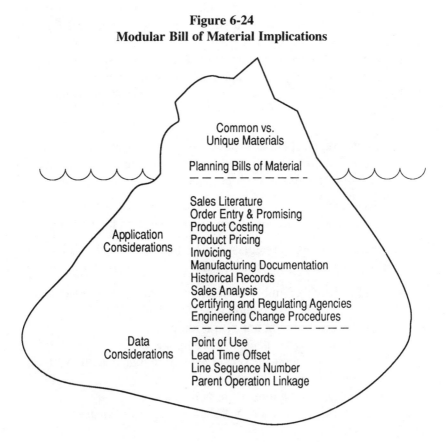

Companies must also consider the effect that modularization will have on historical records, such as those for warranty and traceability purposes. The impact on manufacturing documentation can be extremely important. Outside regulatory agencies impose certification and other regulations that can impact manufacturing records. Sales analysis and engineering change procedures may be affected as well.

It is important that each company address all areas before modularizing. Modularization is not for every company. In many environments—like the iceberg—the impact at the tip is clear. What goes unseen, however, are the many relationships that are indirectly affected by the choice of bill of material architecture.

MAKING IT HAPPEN

With these caveats in mind, companies should consider modularizing their bills of material if their cumulative manufacturing lead time exceeds the acceptable customer delivery lead time and it is impractical to plan and forecast at the end item level. A modularized bill of material architecture can be the perfect approach to the problem.

There is a recommended series of steps to follow when modularizing bills of material:

1. Select a representative product from each family, and layout or depict the time-phased manufacturing/procurement process for the multiple levels of each product.
2. Discuss and understand *where to meet the customer* (competitive delivery lead time versus cumulative manufacturing lead time).
3. Decide stocking strategies—make-to-stock, make-to-order, finish- or assemble-to-order, or engineer-to-order.
4. For finish- or assemble-to-order products, develop modularized bills of material and planning bills of material:
   a. Determine common parts or components.
   b. Determine option-sensitive components.
   c. Identify components sensitive to more than one option. Evaluate redesign possibilities to eliminate.
   d. Develop option forecast to percentages.
   e. Identify opportunities to "flatten" the bill of material.

5. Establish planning bill maintenance as a joint responsibility of planning (master scheduling) and sales and marketing.
6. Develop the order entry process—menu selection, etc.
7. Ensure the integration of planning/master bills and manufacturing bills into one company bill of material data base.

Chapter 7

New Product Introduction and Custom Manufacturing

Larry, a recent retiree from the CCS Carpet Cleaner Company, reminisced over a leisurely meal at the Hunt House one evening:

> "New product launch plans at CCS used to be a joke. Oh, sure, everyone initially would get excited, and there would be a lot of enthusiasm. Early on, we would slip bulletins to the marketplace and advise our stockholders of the newer capabilities and exciting features of our upcoming product release.
>
> "*Upcoming* was probably a good word to describe our continuing dilemma with new product releases. They were always coming, but they never seemed to get there. Probably the most important person in this process was the commercial artist who was hired just before each of the major trade shows. We always seemed to be able to launch an *artist's rendition* of the new product, but never the product itself.
>
> "Eventually, our product design and R&D people would have something to pass to manufacturing. This handoff was never very smooth. For one thing, the pressure on the design group to get something—anything—to manufacturing often meant that the design hadn't really been completed, and the drawings were incomplete, or simply missing, and major parts of the design were still in process or under revision. The marketing department, which was finally seeing some structure to the product design, was usually calling for more enhancements and even for major redesigns based on new market information gathered since the original—some would say ancient—design specifications were signed off.

"But the pressure was on. Design had 'done their thing,' and even though there was some cleanup to complete, it was time to start initial production. At CCS, this would begin with a one-unit prototype from which a manufacturing process and bills of material, routings, and other data requirements could be determined. Sounds easy enough, but guess what? It was at this stage that we would find that the design simply didn't work. To be able to manufacture, heavy redesign was usually required. This sudden realization—it happened almost every time—combined with the fact that the prototype was built on partial design releases and constant design revision, often stopped the prototype and sent us back to design altogether. One time, our chemists released a formulation for a new cleaning solution that worked perfectly every time when 20-gallon batches were made in a laboratory process. In the manufacturing prototype process, however, anything over 100 gallons caused coagulation. The problem, a heating issue, wasn't understood or corrected for three months.

"After several attempts with a prototype, someone would finally dictate that we move ahead to a preproduction run. It's easy to imagine how smooth the preproduction cycle went. Since designs were never stable, engineering changes were issued at incredible rates, and the right materials, which were almost anyone's guess, were never available. To add to this, obsolete materials continued to remain on order, and yesterday's manufacturing process wouldn't support today's revised configuration. The data base, of course, was in shambles.

"Once some level of stability was gained, the manufacturing group frequently saw the opportunity for significant process improvement, which often required even more design modification, but the clock had run out. Unfortunately, through this entire process, sales was taking orders and committing delivery—and recommitting delivery—and we were now losing some orders. Our mode was, Hang the manufacturing expense! Hang missed specifications! Hang reject levels! Hang low productivity! It's off to market we go!

"One ambitious new engineer, new to the CCS product launch process, almost got fired when he identified an $18-per-unit savings with some minor redesign. Unfortunately, the redesign would have added a month of product launch delay! One senior manager even questioned the engineer's basic integrity. Didn't the engineer realize the customer was waiting?

"Of course, the actual market introduction was done with lots of excitement. Parties were thrown. Full-page announcements were made. And we prepared for the next phase of our new product launch scenario—the postlaunch corrective action that makes the design right after the fact. In between, we assessed the blame, identified the required number of scapegoats, and promoted just the right number of the uninvolved, absolutely convinced that such a disaster would never happen again.

"The postproduct launch was sometimes as exciting as the initial launch. Engineering request after engineering request piled up as all the design people tried to clean up the really creative design they never had time to achieve. The manufacturing people tried to implement process improvement, the quality people tried to eliminate the reject causes, and marketing tried to broaden the market share with just one more feature or capability.

"What was really incredible was that no one really seemed to get terribly upset about the whole thing. Finance occasionally pointed out that profit objectives were seldom achieved. Marketing would complain that the product was close to obsolete when it hit the marketplace. The company's practice of slapping impossible cost reductions on manufacturing and purchasing always had some effect, but we never got to the original profitability goals. The biggest reason no one seemed to care about postproduct launch failures was that we were always into the next launch cycle with its typical high level of enthusiasm and expectation. Then one day it happened. One of our competitors hit the marketplace with a product that we planned to introduce in about a year, at a price we thought would approximately be our cost. As you might guess, this got our attention!"

Fortunately for those still working at CCS, what Larry remembers has changed for the better. Today, the CCS product launch cycle is measured in weeks, not years. All design specifications are met, and no engineering change is allowed except to meet regulatory or special customer requirements. This didn't happen overnight. The new product launch process evolved through a series of improvement cycles and continues to improve with every product launch they undertake.

New Product Planning

After CCS's rude awakening with their competitor's new product introduction, a series of senior management meetings was conducted that resulted in two critical decisions. One was to consolidate three new programs into one launch, going only with one new design that they were fairly sure they could achieve. The traditional launch process would be utilized for this product. The second decision was to launch a second new product with a different process—one designed to get the product out faster and reduce their product launch expense. That was the stated mission that led to significant improvement.

At this time, CCS had implemented a fairly successful Manufacturing Resource Planning (MRP II) capability in the company and had significantly improved their basic ability to plan and execute schedules. They decided to apply the concepts and techniques of MRP II to their product launch process.

The first step was to agree that new products would be discussed as part of the sales and operations planning process. They decided to incorporate new products, which were always considered a separate planning issue, into their regular, ongoing management process.

The second step was also to plan each new product launch in the master schedule. Their next new product was an upgrade of the #0123 standard carpet cleaner and was assigned the end item identifier #1123. Based on the sales & operations planning discussions, the master scheduler was asked to plan the phase-in of the new product over a four-month window, phasing out the old product in the same time frame. Any time the new product schedule changed, the master scheduler was told to recommend necessary changes to the old product coverage plan.

Sales and operations planning set the #1123 launch date for June. The master scheduler loaded the following plan into the MPS system:

	May	June	July	Aug.	Sept.	Oct.
#0123 (old)	1,000	800	600	350	0	0
#1123 (new)	0	200	400	650	1,000	1,000

The next step was to utilize material requirements planning to develop and plan component part support for the old and new product schedules. But what components? They didn't even have a bill of material!

To solve this problem, it was decided to assign a couple of people to the design process to generate a material structure and to maintain control of the structure. After some discussion, a team was formed consisting of four members: (1) a design engineer who would act as focal point to the team on all design communication; (2) a manufacturing engineer who was experienced in the company's manufacturing process; (3) a planning specialist who understood the manufacturing systems, the data base requirements, and the planning and scheduling processes; and (4) a purchasing representative to act as a focal point for all new material acquisition requirements (see Figure 7-1).

When this team approach was first announced, the design group expressed concern. They saw engineering design as a creative process, and having a bunch of manufacturing folks around could easily constrain design with too much procedure and control. To resolve this issue,

Figure 7-1
New Product Launch Team

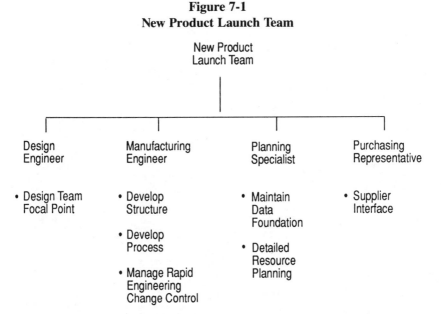

it was determined that the design handoff and ongoing formal definition would come with the issue of an engineering parts list and a preliminary drawing or specification. Any changes after the issue of the parts lists and preliminary drawings would go through a *rapid engineering change control* process (see Figure 7-2).

The rapid engineering control process allowed the new product launch team to evaluate each change, determine its cost, and, if approved, immediately execute its introduction. Nonteam personnel and suppliers were contacted only when team members felt that additional information was required to evaluate or execute the change recommendation. Most changes were evaluated and approved or rejected, and the data files were updated in a matter of minutes. More difficult changes—a tooling modification, for instance—took a little longer, but someone on the team would expedite the information necessary to reach a quick decision and direction.

Once this launch team approach was determined, they returned to the question of component planning: How do you create a bill of material for a product not yet defined? The answer came from the planning people, who suggested that the team set up a *representative bill of material.* Such a bill would be similar to the bill for the current product, but would utilize *dummy* item numbers for new design elements. The #1123 new

Figure 7-2
Engineering/Manufacturing Handoff

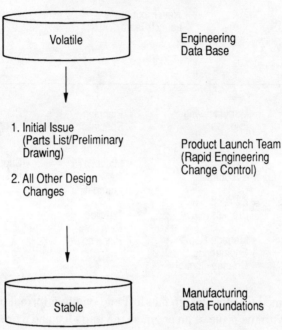

product was an upgrade of the #0123 standard unit with changes in the tank assembly and housing assembly and a new cleaning concentrate. Working with the design engineer, the manufacturing engineer and planning specialist built the multilevel representative structure shown in Figure 7-3.

With the representative bill of material loaded to the company data files, it became possible to plan any existing materials needed for the new product and, more importantly, both to plan the engineering release of new materials needed and to estimate future capacity and labor needs. With estimated lead times applied to the item master file, the representative bill of material shows that the earliest design issue needs to be elements of the housing assembly (see Figure 7-4).

At this point, the launch team came up with an outstanding idea that has greatly contributed to CCS's understanding of new product launch time-phased needs. Working together, the design engineer and the man-

Figure 7-3
Representative Bill of Material

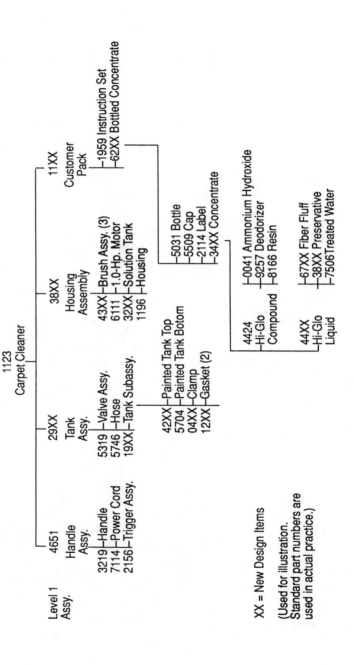

Figure 7-4
New Product Lead Time

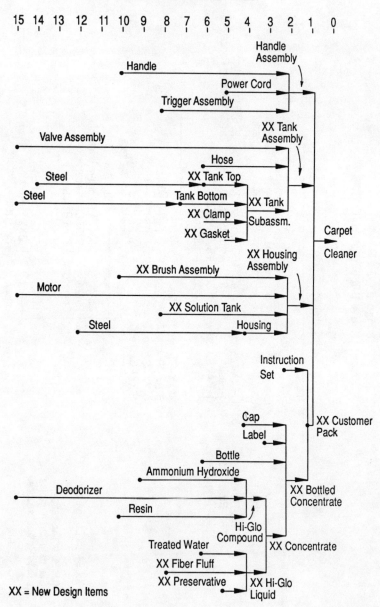

ufacturing engineer created a *bill of activity* by adding tasks, activities, and events to the representative new product structure (see Figure 7-5). It then became possible to use the material requirements planning system to plan all aspects of a new product launch. The material requirements planning process alerted the launch team to any activity necessary to meet milestones and also recognized project *need date/due date* exceptions as dates were rescheduled. This capability caused a lot of excitement when the company began the new #1123 carpet cleaner project, and it has become a vital tool as their product launch control system has expanded to include all new launch programs.

Previous project tracking systems were excellent for individual projects. The problem at CCS was that several projects occurred simultaneously and vied for the same resources. Using material requirements

Figure 7-5
"Bill of Activity"

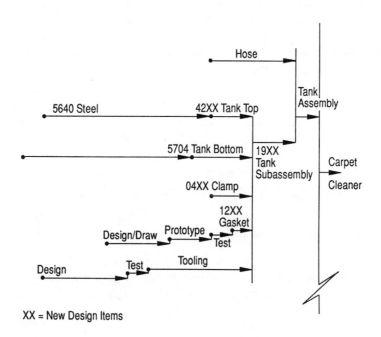

planning and bills of activity allowed the company to employ a single system that included all projects and tied program due dates, material needs, and all other requirements into one time-phased priority planning capability.

The representative bill of material evolved into the actual bill of material as the design progressed. As specifications were developed, actual item numbers were substituted for dummy item numbers.

This structure helped improve the company's priorities for materials in another way. The dummy items with estimated lead times established need dates for lower-level items. Prior to the use of representative bills, design would order in advance certain longer lead time items. Because there were no requirements in the formal system for these items, however, material requirements planning would generate *cancel* recommendations for the advance orders, which caused confusion.

It's interesting to see what happens when new ideas and teamwork develop better ways of managing the way a company does business. Just when everyone thought that everything was covered, the planning specialist pointed out that if they were going to do material and activity planning, why not also load representative routings and do capacity planning? As parts lists were issued and the manufacturing processes determined, the actual routings would replace the representative routings, just as the actual bill structures were to replace the representative bills in the data files. Routing steps were also identified for various tasks and activities on the bills of activities, so that CCS could project resource requirements for design engineering, drafting, and purchasing, as well as for the manufacturing areas.

At this point, CCS launched the new product introduction process using the new model #1123 as their pilot. The process can be summarized by the following steps:

1. Form a new product launch team; might vary with each launch, but typically would have four to six members.

2. Manage the launch at the senior management sales & operations planning level and at the master schedule level using realistic dates.

3. Create representative data foundations, and evolve them to actual product and process definitions as the design stabilizes. Use bills of activities and activity routings to plan nonmaterial launch requirements.

4. Use the planning and control functions of MRP II to achieve schedules for all launch activities and to capture and maintain new product financials.

The #1123 launch marked a significant departure from CCS's traditional launch experience. Everyone knew what had to be done, when it had to be done, how much resource was required, and when that resource was required. Responsibility and accountability were clear, and management was in control.

The company didn't save as much time as they'd hoped on the #1123 because they were learning as they went. There were also a lot of skeptics to convert and a lot of old, bad habits to break. In subsequent new product launches, however, they began to get product to market a lot faster and with a lot less pain.

They also found the next problem they needed to attack. Although they had gained control of priorities and schedules, they were still rescheduling, because of significant problems manufacturing the product after it was handed over from design.

DESIGN FOR ASSEMBLY

The problem of having a product design that did not fit CCS's manufacturing and assembly process was eventually solved. First, however, they needed to recognize just how serious a problem this was. Once they were operating in a more formal manner with better control of their schedules, the reasons why they constantly rescheduled became painfully obvious. Their biggest scheduling headache was the continual redesign needed to achieve a prototype that could then be moved to a preproduction manufacturing run. Costs kept escalating, especially the expense of postlaunch engineering changes. There was no question that something had to be done.

The solution came from the quality staff, who had heard about *Design for Assembly (DFA)* software and had made some initial inquiries. From these inquiries they developed a pretty strong case that the Design for Assembly process was providing real benefits to many companies. It appeared that it could help CCS in the early detection of assembly problems in their new product designs.

Currently, there are several DFA systems (some software based, others noncomputer driven) that allow for a quantitative analysis of a product design. DFA simplifies the structure of a product by eliminating or combining parts to simplify the assembly process. The analysis quantifies the design on the basis of the number of parts and the ease of assembly.

After a brief review of the packages available that fit CCS's needs, the company selected a software-based capability and then picked and trained a team to implement the design review on a series of modifications being introduced on the domestic line of heavy-duty cleaners. The team consisted of a design engineer, a manufacturing engineer, and a quality engineer. They then began to implement the Design for Assembly process.

At first, there were some problems with design engineering because, in effect, the process was questioning, on the basis of assembly capability and numbers of parts used, the basic design of the product after the fact. Redesign always took time, and the pressure to "get it out the door" often prevailed. But everyone was well aware of the alternatives—to redesign it later at the prototype stage or to launch it with excessive manufacturing cost and inefficiency.

The results have been very impressive. CCS's newer products contain considerably fewer parts, cutting overhead in ways that were never imagined (see Figure 7-6). But most importantly, their product launch time, manufacturing costs, and postlaunch engineering changes have decreased significantly, while product quality and manufacturing productivity have substantially improved.

Nice story? Happy ending? Not quite! CCS is now operating at a class A level. It is company policy to be totally discontent with the status quo. Their product launch practices have continually improved. Today, at CCS, they *design for manufacturability* or, as our Oliver Wight associate Dale Hiatt likes to say, "They design for competitive advantage."

DESIGN FOR MANUFACTURABILITY

The design for manufacturability process considers the requirements of manufacturing right from the initial design concepts. The process calls for teamwork from the word go, with joint design/manufacturing cooperation at every phase of the product launch process.

Design is not handed over to an early product launch team. The manufacturing people are part of the design team, and manufacturing data requirements are stabilized as part of the design process. The

**Figure 7-6
Eliminated Parts Are Never . . .**

□ Maintained in Documentation Control
□ Ordered by Planning and Scheduling
□ Expedited Because They Are Late
□ Received
□ Inspected
□ Dispositioned When Discrepant
□ Held in Inventory
□ Damaged by Handling
□ Obsolete and Written Off

---

company data files are electronically maintained (see Figure 7-7). *Concurrent* or *simultaneous* engineering are similar efforts intended to improve and speed the design process.

As the product moves closer to prototype and premanufacturing, more and more manufacturing personnel join the effort. In essence, each product launch team is a product company or product division. The

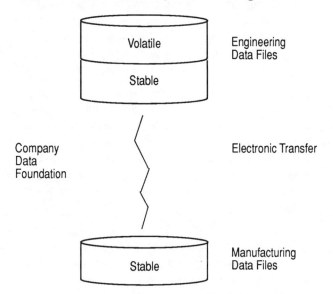

**Figure 7-7
Design for Manufacturability—Data Management**

team exists from the first design phase start-up until the product is discontinued as a manufactured item. The makeup of the team changes, but the overall management, including a senior design engineer and some design staff, remain on through the entire product life cycle.

CCS has formally identified each major step of the design phase and the specific deliverables of each of those steps (see Figure 7-8). A launch team never releases a product from one phase to the next until every requirement or specification is achieved. This includes new product design objectives as well as the manufacturability of the new design. If they get the design right the first time, they never have to redesign. If a problem arises during one of the design phases, the problem-solving tools of Total Quality Control are brought to bear. Every member of the launch team contributes to the problem-solving process. Each phase of the launch is scheduled, and the overall schedule is driven by meeting each phase plan.

CCS continues to use the Design for Assembly process, but it's never done after the fact now. The joint launch team makes checks at every step of the process. Parts do not have to be eliminated from the design, however, as part simplification was considered right from the start. Manufacturing process improvement is not a follow-up activity; the best manufacturing process is considered in the initial design.

New ideas surface, of course, as the new product moves from concept to market introduction. Such ideas are always encouraged, but they are never incorporated into the current new product launch specification unless (1) they can be incorporated into the design at its current phase without returning to a previous phase, and (2) the current phase schedule is not impacted. For example, a better idea that required a drawing change would not be allowed if issued drawings needed to be revised. Likewise—and here's a real departure from tradition—engineering changes are not allowed, except those dictated by regulation, safety, or customer demand, through the entire product life cycle!

But what about all the good ideas? It's simple. They will all be incorporated into the next product launch, or if the next launch is already in progress, the one following that. At CCS, product launch has become a routine element of the business process. The specs are set, the specs are met, and the specs aren't changed. The requirements of the customer and of manufacturing are achieved rather than traded off against each other. Schedules are also met, and total quality is attained at every step of the product launch cycle.

The CCS Carpet Cleaner Company is, of course, a fictitious company

Figure 7-8
Product Launch—Phases and Deliverables

Phase	Deliverable
Concept	Design Requirements
Product	Product Definition
Design	Drawings & Data Set
Acquisition	Material, Tooling, Gauges, Etc.
Production	Prototype/Premanufacturing/Actual Production Run
Launch	Announcement & Sales
Maintain	Product Life Support

created to share concepts, principles, techniques, and ideas with the reader. But nowhere are the capabilities of the CCS Carpet Cleaner Company outside the realm of what is happening in industry today.

Where you begin is determined by where you are. Using representative bills of material and bills of activity, with representative routings, to support time-phased planning of product launch programs provides an excellent starting point. if your company has successfully implemented MRP II software and practices, the tools you need are already in place. The Design for Assembly processes are readily available and relatively inexpensive to acquire. The major issue will, of course, be the ability of your design and manufacturing organizations to work together. As always, people are the key. The method to accomplish this is still the same: education, discussion, and pilot implementation.

The Design for Manufacturability process is typically implemented by companies that have successfully installed Just-in-Time and Total Quality Control programs. Design for Manufacturability follows the basic philosophy of eliminating waste and encouraging teamwork and formal problem-solving methods. If Just-in-Time and Total Quality Control are part of your company's operating practice, the process of Design for Manufacturability is probably already in place or being implemented. If not, it deserves serious consideration. Design for Manufacturability education would be the first step in implementing such a new product launch capability.

CUSTOM PRODUCT MANUFACTURING

In some businesses, every unit sold requires a design effort, and the company is basically in a continuous new product launch mode. Often labeled *engineer-to-order*, such companies begin with a customer specification, move through a design process sometimes called *application engineering* or *customer engineering*, and finish with a product delivery. Often, a single product is all that is delivered; sometimes a few are made.

The amount and nature of design/development work required in application engineering or customer engineering varies considerably from company to company. For some, the design effort is an extensive process involving significant technical issues, customer review, and approval cycles. For others, it's much more routine and may be performed by less technically trained individuals found in customer order entry or cost-estimating groups.

The nature of the business can make it difficult to do any planning. Bills of material and routings are at least partially undefined, so lead times and capacity requirements are difficult to determine. Order promising is therefore also difficult, as are estimating cost and quoting price.

Many companies producing custom products have, however, implemented excellent planning and control systems in spite of these unknown or uncertain factors in their business. To accomplish these capabilities, they have utilized MRP II systems and product launch processes to make maximum use of the known elements of their particular business.

Typically, a company has a set of sales and field support expertise, engineering skill, and manufacturing capability organized to deliver a given product to the marketplace identified as a company's product families or models. For example, a manufacturer of custom turbines may define that company's families as large and standard gas powered, large and standard diesel powered, and steam powered. A custom truck manufacturer, on the other hand, may offer three families based on gross vehicle weight of the chassis.

One common approach is to create a *generic bill of material* and a *generic routing* for each family or model. Supporting item information represents the usual material content and usual manufacturing processes that a typical custom product requires (see Figures 7-9 and 7-10). These generic data foundations are developed and updated based on the history of a product family's requirements and on the knowledge of the personnel involved in designing and manufacturing that specific family of customized end items. Average costs for materials and manufacturing operations are frequently tracked based on actual cost reporting. Job costing systems are often used.

Initial and preliminary quotes are often prepared by comparing customer requirements against the generic bills and routings to determine major design variances. Costs are added or subtracted accordingly. More detailed quotations are accomplished with more finite discussion and review. Delivery dates are determined by looking at current customer order load and adding the estimated load generated by the new customer requirement.

When an actual order is received, the generic data for a product family is copied and then corrected to the specific design requirements of the customer. Data such as lead times and costs are also corrected based on the actual bill of material and routing developed for that customer order. *Generic* bills and routings are very similar to the *representative* bills of material described earlier.

Figure 7-9
"Generic" Bill of Material

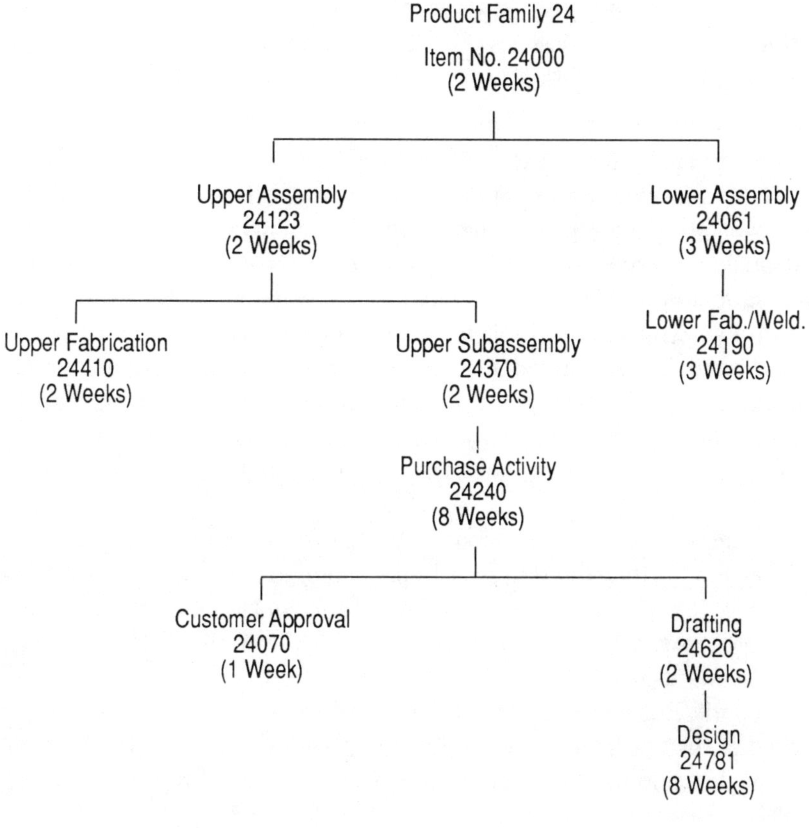

```
                    Product Family 24
                     Item No. 24000
                      (2 Weeks)
        ┌─────────────────┴─────────────────┐
   Upper Assembly                      Lower Assembly
      24123                                24061
    (2 Weeks)                            (3 Weeks)
        │                                    │
        │                             Lower Fab./Weld.
        │                                  24190
        │                                (3 Weeks)
   ┌────┴─────┐
Upper Fabrication    Upper Subassembly
    24410                 24370
  (2 Weeks)             (2 Weeks)
                            │
                     Purchase Activity
                          24240
                        (8 Weeks)
                ┌───────────┴───────────┐
          Customer Approval          Drafting
              24070                   24620
            (1 Week)                (2 Weeks)
                                        │
                                      Design
                                      24781
                                    (8 Weeks)
```

Figure 7-10
"Generic" Routing

PART NO.	OP. NO.	DESCRIPTION	QTY.	EST. HRS.	WORK CLUSTER
24000	10	Assemble & Test	1	24	Assy.-60
24061	10	Prep. Routing	1	5	Meng.-80
	20	Assemble	1	68	Assy.-50
24070	10	Customer Approval	1	10	Sales-30
24123	10	Prep. Routing	1	5	Meng.-80
	20	Assemble	1	36	Assy.-50
24190	10	Prep. Routing	1	5	Meng.-80
	20	Fabrication	1	34	Fab.-40
	30	Weld.	1	21	Weld.-20
24240	10	Purchase Material	1	45	Pur.-10
24370	10	Prep. Routing	1	5	Meng.-30
	20	Subassemble	1	29	Assy.-70
24410	10	Prep. Routing	1	10	Meng.-80
	20	Fabrication	1	38	Fab.-40
24620	10	Drawings	1	24	Drfp.-80
24781	10	Design Complete	1	165	Deng.-91

Forward planning and detail scheduling are accomplished using MRP II. Sales and operations planning, master production scheduling, rough-cut capacity planning, and material requirements planning ensure that adequate resources are available and that valid schedules are released to the suppliers and the plant. Detail capacity planning and shop floor control systems are used for plant scheduling if required.

Demand is determined in the short term by the customer orders (backlog or order book) and the definition of the manufacturing requirements to support those orders. Demand over the longer term is determined by a sales forecast of anticipated orders by product family.

Figure 7-11 illustrates a company plan for a family of large capital equipment further divided into light, standard, and heavy-duty. Customer orders are numbered (e.g., #L-148), and forecasted orders are indicated as generic structures (e.g., #L-FCST). Typically, three lights, two standards, and one heavy-duty customized product are scheduled per month. Because of a special order for heavy-duty units, however, the schedule has been adjusted slightly in periods 2 and 3. The vertical segmented line indicates the customer engineering and manufacturing lead time. Actual orders have not materialized for standards in period 4, but because sales has indicated that this is a short-term problem, management has decided to move future orders forward to fill the capacity void in that period.

When asked how successful they are at getting design data to manufacturing on time, many managers from custom product companies will simply chuckle and suggest that such a thing simply doesn't happen in the real world. One person, when asked when drawings are issued, responded: "They're not. We mail them to the customer just after product shipment, hopefully just before product delivery. We work with pencil drawings and verbal instructions. Of course, we do a lot of rework, and occasionally the customer doesn't quite get what they expected!"

Many manufacturers of custom products have solved this dilemma by including design activities in the product family generic bill of material, along with routings and the estimated hours for those activities. This allows the company not only to resource and priority plan material and manufacturing capacity, but also to plan critical engineering, field sup-

Figure 7-11
Custom Product—Production Plan

	Product			Production Plan		
1	2	3	4	5	6 $\longrightarrow$
L-136	L-144	L-148	L-156	L-163	L-FCST
L-140	L-145	L-149	L-159	L-FCST	L-FCST
L-143		L-153	L-160	L-FCST	L-FCST
			L-162		
S-137	S-142	S-147	S-FCST*	S-157*	S-158*
S-141		S-150	S-FCST*	S-FCST	S-FCST
		S-152			
H-138	H-139-A	H-146	H-151	H-161	H-FCST
	H-139-B				

Design & Manufacture:
Lead Time

* Action determined: move future orders
to fill capacity void—incur storage cost
of stored finished product.

L = Light
S = Standard
H = Heavy

port, and customer interface. Again, this technique is very similar in practice to the bill of activity used for new product introduction.

The unknowns of custom product manufacturing can be planned by building on the known, creating the product family generic bills of materials and routing structures, and then using the power of MRP II to manage priorities and plan resource capacities. Such a system supports the longer-range, aggregate sales planning, manufacturing capacity planning, and financial planning of a company. It also provides data for an effective sales and operations planning process. Such a system facilitates quoting orders, delivery promising, and customer design detail definition.

MAKING IT HAPPEN

Using company data foundation and MRP II techniques to assist in the planning and control of new product launch may be a major departure from current company practice. We suggest, however, that some of the techniques discussed in this chapter might be beneficial in improving product launch success. Because such approaches may be very different from current practices, we recommend a small pilot start-up in which a launch team, following education, would determine which approaches or techniques outlined in this chapter would be implemented in support of the next product launch activity. This team would then become responsible for the management of the next launch and would apply and debug the techniques selected as they proceeded. Once debugged, the new methodologies would be expanded to all future launch programs. This pilot process provides a fail-safe start-up and all the success of the pilot to justify continuing the new approach for future launch activities.

Similar approaches can also be used to plan and schedule custom product or engineer-to-order manufacturing. They, too, should be initiated on a pilot basis, perhaps by selecting a single-family product. The steps involve the education of a small task force representing all the functions that would be involved in using generic data files; the development of those generic files for the pilot family; and the determination of how those files would support the planning, quoting, and customer configuration requirements within the existing planning and control systems. Once one product family has been tested and implemented, the balance of the families would have their generic data files determined and loaded to support full company use of the improved planning and control capability.

Once a product is launched or is designed to a customer specification, the design problems need to be managed. The management of engineering change control is the subject of our next chapter.

Chapter 8

Managing Engineering Change Control

Today's manufacturing community is strongly focused on continuous improvement and the elimination of waste. This "One Less at a Time" process promotes small but increasing numbers of changes to our products and processes. Since it is an approach that fosters change, it also forces us to address our ability to cope with change.

To deal with this increasing amount of change, new and more responsive methods of managing and implementing change are required. Continuous improvement also requires new attitudes, behavior change, and a willingness to allow for change control at the supervisory and production personnel level.

Although encouraging change, we create a risky situation if we are not in a good position to manage change properly. The curse of uncontrolled changes leads directly to loss of market opportunity, obsolete inventory, material shortages, and manufacturing inefficiencies. Ill-managed change can also cause missed deliveries, poor quality, and loss of configuration control. This, in turn, generates frustration in the ranks, low morale, and poor productivity. These maladies will invariably translate into a series of missed opportunities for a company. In the highly competitive global arena, few businesses can afford to risk the possibility of not improving their market position.

This chapter covers two areas of engineering change control. The first involves the people issue of policy and administration. The second involves the data base and computer systems available to control new material replacement or cut-in control.

ELEMENTS OF GOOD ENGINEERING CHANGE POLICY AND PROCEDURE

At CCS, the need for a comprehensive engineering change process has always been recognized as an important element of business success. At the policy level, the management team established an engineering change review board, an engineering change coordinator, an engineering change review process, and appropriate performance measures.

The permanent members of the engineering change review board are the engineering change coordinator and representatives from design engineering, manufacturing engineering, material planning, purchasing, finance, and manufacturing. Ad hoc members include representatives from quality control, sales and marketing, safety, and field service.

The CCS review board is responsible for overseeing both the engineering change request (ECR) and the engineering change notice (ECN) process. The board meets at regular intervals to discuss and dispose of change requests and to review and initiate corrective actions on approved changes that are behind schedule. The length of these regular meetings varies, depending on the number and critical nature of the changes to be discussed. Like most companies, CCS finds that a regularly scheduled weekly meeting helps to keep their engineering change process responsive.

The person in charge of this team is the engineering change coordinator, who is responsible for coordinating the effective implementation of engineering changes. This means that the engineering change coordinator must be someone who is able to work closely with design, manufacturing, planning, and purchasing. This person must also be able to verify that the information regarding the change is added to the company data foundations in a correct and timely manner.

CCS's change policy states that anyone is allowed to initiate an engineering change request. Such requests are documented on an originating request form on which the suggested change, impact, and benefit are described (see Figure 8-1).

The CCS engineering change policy establishes the authority levels for approving change. Most change requests are, of course, authorized by the board itself after an evaluation process, which we will discuss shortly. Normally, management's approval is necessary when issues such as significant cost, product impact, and capital outlay are involved. There are also special cases when a change might cause an environmen-

Figure 8-1
Request for Engineering Change

Date: _____
Originator: _____
Dept.: _____
Product(s): _____

Suggested Change:

Any Impact:

Expected Benefit:

Authorization: _____
(For Production Improvement)

Authorization: _____
(Required Only If Mandatory)

Engineer Assigned: _____

ECR Number Assigned: _____

---

tal impact. In these situations, the environmental manager has to give approval. Similarly, if a change affects a government regulation or safety standard, the approval of management and sometimes an outside agency is required. At CCS, the engineering change coordinator is allowed to approve obvious changes without board approval, provided these changes do not exceed a specific cost threshold.

Another area under the board's domain is the actual change review procedure itself. Every originating request is logged in when received. The engineering change coordinator evaluates and accepts some requests immediately and forwards them to the responsible engineer to prepare the engineering change notice and any supporting information, such as drawing revisions. Figure 8-2 shows a typical engineering change notice.

When the requested change requires board review prior to approval, the engineering change coordinator, after logging in the change, assigns the request to the responsible engineer to prepare a formal engineering change request for distribution and impact analysis (see Figure 8-3). (Note that we use the term *engineering change request* as a preapproval control document and *engineering change notice* as a postapproval document.)

For example, industrial engineering is responsible for analyzing the changes affecting process, tooling, operator training, routings, and material handling. The materials department is responsible for an analysis of obsolete materials and lead times for new materials. Quality assurance is responsible for assessing any quality processes that are impacted. Finance looks at the impact on product cost and profit. This process may be referred to as the *engineering change request distribution*.

The time allowed for this process is specified by policy. CCS allows five working days for normal priority requests. If a change needs to be processed faster, there are special procedures to "walk through" such *emergency* change requests in twenty-four hours. CCS established these emergency change conditions carefully, so as to avoid the possibility that every change would be regarded as an emergency.

To speed up acceptance of plant recommendations for changes that would improve production processes, the responsible supervisor and the engineering change coordinator have approval authorization. If approved, an engineering change notice is immediately issued.

There are times, of course, when mandatory changes are dictated by regulation, customers, or management direction. Rules have to be defined to handle these procedures. If a change is mandatory, the engineering change request distribution is still necessary to assess implementation requirements and cost, even though the board does not determine acceptance.

Most of CCS's change requests flow smoothly and quickly. When a potential rejection does arise in the process, disputed issues are scheduled to be addressed at the next change review board meeting.

For approved changes, there is a final distribution that notifies each department within CCS of approved changes and triggers whatever is necessary to initiate the change.

Figure 8-2
Sample ECN

ECN No.: 3462	**Notification Given to:**
Change Class: Requested ___ Final ✓	Administrator ☑
By: RAL	Design ☑
Date: 2/8	Data Processing ☑
	Sales ☑
Part No. & New ECN Letter:	Accounting ☑
3804 Housing Assembly ; 7612 ;	Mfg. Engr. ☑
9550 ; 4421 ; 6507 ; 7203 ; 8461	Assembly ☑
	Fabrication ☑
	Chemical ☑
	Purchasing ☑
	Quality ☑

Description of Change: Supercede 1196 Housing (metal) with 2178 Housing (Plastic)

Comments: Use up existing stocks.

Effectivity Date: Requested 2/1
Finalized 2/8
Completed 3/24

Approved by: DOC
Date: 2/8

Figure 8-3
ECR Distribution

Production Control
    Qty. on Hand: _____   Qty. of Rework: _____
    Qty. W.I.P.: _____   Qty. to Scrap: _____
    _____   Avail. Date: _____
    _____   If Scrap: $ _____
    P.O. Qty.: _____   Rework: $ _____
    Total Qty.: _____   Lead Time: _____

Mfg. Eng.
    Cost of
      New Tooling: _____   Labor Std. Change:
                                  Yes _____ No _____
    Cost of Labor   Routing Change:
      to Implement: _____   Yes _____ No _____

Purchasing
    Cost: _____   Lead Time: _____
    Vendor: _____

Quality
    Requirements: _____
    Cost: _____

Design
    Drawing Change: New Part No. Req.:
      Yes _____ No _____   Yes _____ No _____
    B/M Change: Order No.: _____
      Yes _____ No _____

Accounting
    Cost: _____

Sales
    Warehouse No.: _____   Rework or Scrap: _____
    Qty. On Hand: _____   Order Entry Change: _____
    Send Cust. Prints:
      Yes _____ No: _____

Managing Engineering Change Control

CCS has found it helpful to have a tracking process in place, maintained by their engineering change coordinator. An engineering change notice in dispute, behind schedule, or facing a schedule slip becomes an agenda item for the next change review board meeting. It is the board's responsibility to review the impact of the slip or delay and take corrective action. This proactive review of delayed engineering change notices allows for corrective action to be taken, rather than waiting and discovering after the fact that a change has failed to accomplish its objective. It also helps to keep the process in focus.

UPDATING THE DATA FOUNDATIONS

Processing an engineering change notice usually requires data foundation updates from several organizations. Design often is required to issue drawing revisions, bill of material revisions, or specification updates. This can be selective, since the ECN may change a bill of material but not the drawing or some other piece of controlled information, such as an automatic insertion program. Planning and purchasing may need to add new lead time and lot size data. Manufacturing engineering may require routing revisions. Cost data often must be modified, and special coding (e.g., parent operation, phantom, and engineering status) may require update. The process of updating the company data foundations may occur in three different ways:

1. Each organization is made responsible to enter its own particular information. Finance is responsible for cost data, materials for planning data, design engineering for bill of material data, and manufacturing engineering for process and routing changes. This approach clearly defines responsibility, allows data to be loaded as they are determined, and quickly identifies errors to be resolved. However, this approach only works in a company where the culture calls for a high degree of data integrity. It is essential that everyone trusts everyone else to maintain his or her own data in an accurate and timely fashion. This is the norm in a class A company.
2. A single, central data control function collects all required information and maintains the company data files. This function is sometimes referred to as *configuration control, bill of material administration,* or *data administration.* It audits the information, makes sure it is complete,

and coordinates and expedites data entry to support the engineering change implementation objectives. Problems can occur with this approach. Responsibilities can become unclear. Also, error correction demands coordination of central and source organizations. Some class A companies do, however, utilize the central maintenance approach.

3. A hybrid approach also exists in which some data are source oriented and some are centralized. This approach is often implemented where a bill of material coordinator maintains bill of material information, manufacturing engineering oversees routing information, and each department maintains its own item master information.

All three approaches can handle the successful processing of an engineering change notice. The third, however, is probably the most popular.

With the advent of computer-aided design (CAD) capabilities, many companies have tried to electronically link CAD to the manufacturing data files. This is usually an incomplete solution, because the engineering design does not always reflect manufacturability and because other data may need to be fashioned to support bill of material creation for the manufacturing files (see Figure 8-4).

Some companies, trying to build engineering/manufacturing team-

Figure 8-4
Engineering/Manufacturing Interface

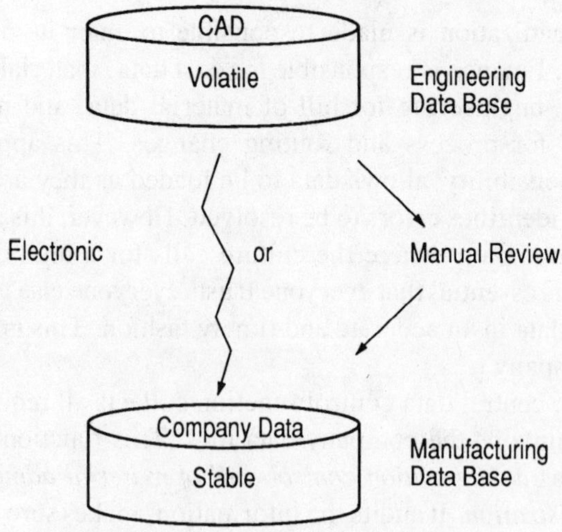

work to improve product launch timing, have solved these problems by building manufacturability into the engineering design. As the design stabilizes, other support data are developed and loaded to the manufacturing data files in order to support down-loaded CAD data. This subject was discussed in greater detail in Chapter 7.

MEASURING PERFORMANCE

An important task of the CCS review board is to develop and maintain the performance measures that monitor their engineering change process. Their policy calls for a designated period for processing change requests. Monitoring their performance also means measuring the number of changes coming in, the number that are being addressed, and those that are in queue waiting to be addressed, as well as those that have been rejected and accepted. This allows the board to measure if the request is being turned around appropriately, while also appraising CCS of the backlog of requests in process (see Figure 8-5).

Where certain benefits are expected, it is important to monitor benefit achievement. Such achievement might be fairly straightforward, such as meeting standards or getting regulatory agency approval, or a little more

**Figure 8-5
Engineering Change Performance**

Engineering Change Requests

A. ECRs Reviewed Year to Date: _____
ECRs Rejected   " " " : _____
ECRs Accepted   " " " : _____

B. Beginning ECR Backlog: _____
New ECRs Received: _____
ECRs Approved: _____
ECRs Rejected: _____
Ending ECR Backlog: _____

C. Aging of ECRs in Process:
<1 Week   _____
1–2 Weeks   _____
2–4 Weeks   _____
5–8 Weeks   _____
>8 Weeks   _____

difficult, such as increased sales. Regardless, all achievement must be measured so that the payback, as well as the cost of engineering change, is identified.

TECHNIQUES FOR IMPLEMENTING ENGINEERING CHANGE

Now that we have examined the people, policy, and administration issues, it's time to focus on the computer-assisted aspects of the phasing in and out of component change. If a change is to occur immediately, the bill of material needs to be instantly updated, with new parts planned without considering the consumption of old parts. However, if existing inventories are to be used up before the new part is cut in, there are three major ways to use bills of material and the material requirements planning system to accomplish a phased component cut-in. Most software packages offer some variation of effectivity techniques and also have the ability to use firm planned orders as an engineering change mechanism. One technique that is not software dependent utilizes an artificial bill of material relationship to *use up* or phase in/phase out items.

Date Effectivity Technique

The date effectivity method, a popular approach for phasing in part change, is available in most MRP systems. The mechanics of this technique are fairly straightforward. The software allows the user to carry a date at which the old part planning ends and the new part planning begins.

The cut-in/cut-out dates are based on the remaining on-hand and on-order supply compared to the current demand for the component being replaced. An earlier set of dates may be selected if only a partial supply of the item being replaced is to be used.

Engineering change notice #3462 calls for CCS to replace the metal housing—item #1196—with a new plastic housing—item #2178. The parent housing assembly's item identification remains #3804, as its form, fit, and function are unaffected.

Two software methods can accomplish this time-phased date effectivity approach. In each approach, item #2178 is added as a component to #3804 without deleting #1196, which is being replaced. The first method places the effectivity dates directly in the bill of material component record. Using this method, a pending delete date is assigned to the metal housing, and a pending add date is assigned to the plastic housing (see Figure 8-6).

Some software allows a third date, called *coverage through*, in which the part being replaced is planned to maintain coverage beyond its cut-out date as protection against new part implementation problems, demand fluctuation, or inventory adjustment.

The second date effectivity method is a slight improvement over the first. The delete and add dates are used, but instead of carrying the effectivity date in the bill of material component records, the software carries an add/delete code tied to a reference number, usually the engineering change number itself. There is also an auxiliary file, which states that the parent item #3804 housing assembly is being impacted by engineering change notice #3462 (see Figure 8-6). The date associated with that change is effective on 3/24. The advantage of this approach is that as demands or other circumstances change, the company can manage the effectivity date in one place without having to go into the individual component records to change it.

As the material requirements planning system performs level-by-level

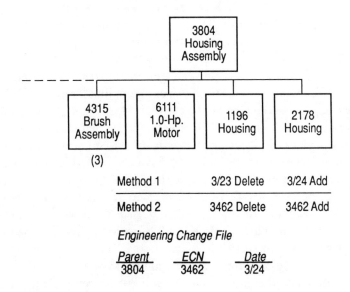

**Figure 8-6
Effectivity Date Method**

planning, the program constantly checks effectivity in the component records. In our example, material requirements planning would plan old housings (#1196) through 3/23 and new housings (#2178) beginning on 3/24 in support of any planned order releases of the parent (see Figure 8-7). The requirements for #1196 through 3/23 are satisfied with remaining inventories, and the requirements for #2178 after 3/24 cause the MRP system to release orders for the replacement part.

We need a few words of caution here. Suppose that as we move forward in time, the demand increases for the #3804 housing assembly (see Figure 8-8) and the increased planned order releases from #3804 explode to the old housing. In the previous example, there was a balanced plan that allowed CCS to run out of stock at the cut-out date of the engineering change. Unfortunately, now the demand increase has triggered MRP to send a message telling the CCS planners that they're going to run out of stock before 3/23, and the system is recommending that they place an order for the old housing. That, however, is not what they want. This is why it's so very important to have some way to indicate to the planners that an item is being phased out under an engineering change. Some companies send a message in a special instructions field. Others may set an item master code that won't allow an item to be reordered. Without a warning system, some morning when the CCS planner has 100 plans to review, and #83 is this recommended order for old housings, the old housings could easily be ordered. It's very easy for these time-phased effectivity changes deep within the bill to get lost in the huge volume with which a planner must deal.

Some repetitive manufacturing companies use a form of the date effectivity technique known as the *block change*. In this particular approach, a series of engineering changes are timed simultaneously and put into effect at one specific time. An excellent illustration of this block change is represented by the model car year. Most years, the new models come out in the fall with all their new and improved changes. In some instances, a block engineering change is done to control configuration. Units produced between certain dates have a specified bill of material, which is recorded in the product's configuration history.

Another method of effectivity change control is the *serial number effectivity technique*. In this approach, a *from* and *to* serial number effectivity is maintained within the bill of material record. The bill of material system thus carries a range of serial numbers from which a component is authorized for use. The planning system selects the proper structure based on the parent item serial numbers being planned.

Managing Engineering Change Control 173

Figure 8-7
Effectivity Date Method—Material Planning

Part No. : **3804**   Lead Time : _____   Order Qty.: _____
  Housing

	PERIOD							
	1	2	3	4	5	6	7	8
Projected Gross Requirements								
Scheduled Receipts								
Projected Available Balance								
Planned Order Release	20	30	15	25	20	30	10	40

Pending Delete Item   3/23

Part No. : **1196**   Lead Time : **2**   Order Qty. : **80**
  Housing

	1	2	3	4	5	6	7	8
Projected Gross Requirements		20	30	15	25			
Scheduled Receipts								
Projected Available Balance	90	70	40	25	0	0	0	0
Planned Order Release								

Pending Add Item   3/24

Part No. : **2178**   Lead Time : **4**   Order Qty. : **100**
  Housing

	1	2	3	4	5	6	7	8
Projected Gross Requirements					20	30	10	40
Scheduled Receipts								
Projected Available Balance	0	0	0	0	0	-20/80	50	40	0
Planned Order Release	*100							

*Exception Notice — Place Order

Figure 8-8
Effectivity Date Method—Increasing Demand

Demand for __3804__ Increases;

Planned Orders for __3804__ Increase

3804 Housing Assembly	PERIOD							
	1	2	3	4	5	6	7	8
Planned Order Release	20	30	15	25	20	30	10	40

1 Week Later

Planned Order Release	✕	30	15	30	25	35	15	50

Pending Delete Item

Part No.: __1196__ Lead Time: __2__ Order Qty.: __80__ (3/23)

Housing

	1	2	3	4	5	6	7	8
Projected Gross Requirements		✕	30	15	30			
Scheduled Receipts		✕						
Projected Available Balance	70	✕	40	25	-5 / 75	75	75	75
Planned Order Release		✕	*80					

*Exception Notice — Place Order

Pending Add Item

Part No.: __2178__ Lead Time: __4__ Order Qty.: __100__ (3/24)

Housing

	1	2	3	4	5	6	7	8
Projected Gross Requirements		✕			25	35	15	50
Scheduled Receipts		✕			100			
Projected Available Balance	0	✕	0	0	0	75	40	25	-25 / 75
Planned Order Release		✕		100				

Managing Engineering Change Control

Serial number effectivity must support two conditions. In the first, upper-level item numbers are changed when form, fit, function, and interchangeability rules require it. In Figure 8-9, the before condition illustrates that the components structured into the parent #0123 are effective for all future production. When an engineering change caused #0127 to replace #0123 beginning with serial #315, two bill of material maintenance activities were required. The first was the addition of the new #0127 and its components to the data base. The component serial range is effective from serial #316 to serial *On*, indicating that all future production of units after serial #316 will be in this new configuration. The second activity was the revision of the serial number range of the components of the #0123 to indicate that serial #315 was the last use of the #0123 configuration.

A second serial number effectivity condition occurs when the parent item retains its item number and one or more components are affected. In Figure 8-10, the engineering change notice calls for the #7411 tank subassembly to be replaced by a #6603 tank subassembly effective with serial #351 production unit of the #0127 parent. In this situation, the only bill of material maintenance required is to add the new #6603 component with a range of serial #351–On and to change the range of the serial #7411 from #316–On to #316–#350. This condition results in multiple serial number ranges for components structured into a given parent. Therefore, the material planning process is more complex when planning component support based on the planned serial number production of the parent item.

Not all companies require the serial number effectivity capability just described. Many manufacturing facilities that need to track serial number configuration will do so on ancillary files while managing engineering change control with date effectivity or firm planned order techniques. Where true serial number effectivity is required, the selection of software with this capability is quite limited and is usually part of a software package designed for aerospace and defense contractors. In this arena, there are typically fewer end items and lower volume, making serial number effectivity more easily managed.

Firm Planned Orders

The firm planned order (FPO) method of new part phase-in planning uses the capacity to set up an order-specific bill of material for one order of an item that differs from the standard bill of material for that item.

176 Manufacturing Data Structures

**Figure 8-9
Serial Number Effectivity—End Item Change**

(Before) Engineering Change Notice

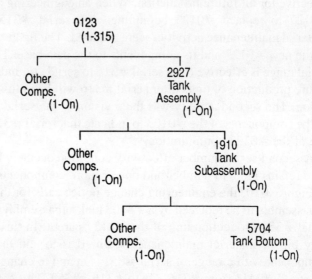

(After) Engineering Change Notice

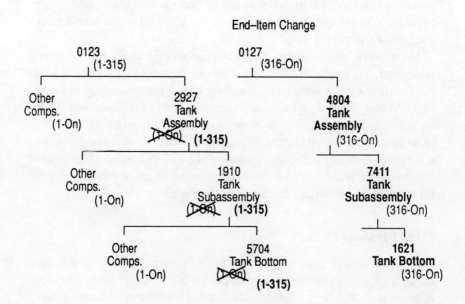

Managing Engineering Change Control 177

**Figure 8-10
Serial Number Effectivity—No End Item Change**

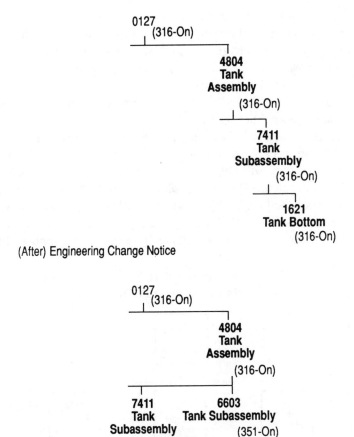

Once the remaining inventory of the old part to be used up is determined, a two-step process is initiated. In step 1, a firm planned order for the parent is established for the quantity of the old part, using the then current bill of material structure. Step 2 involves changing the bill of material to the new part number, and this occurs immediately after the first step. Material requirements planning then plans all future orders for the new bill of material.

178   Manufacturing Data Structures

Let's look at CCS's #4651 handle assembly as an example. Suppose that CCS plans to replace the existing trigger assembly with a newer design, but that change also necessitates a new handle. There is no form, fit, function, or interchangeability issue with the #4651 handle assembly itself (see Figure 8-11).

CCS also wants to maximize the use of existing inventory prior to making their change. Since the most expensive part in their handle assembly is the #2156 trigger assembly, they want to maximize their use of this part. The idea is to try to marry up the #3219 handle inventory accordingly into matched sets.

MRP goes to work. As Figure 8-12 shows, currently, there are 5 #4651 handle assemblies on hand and 75 due in this week. According to the projected available balance line, CCS anticipates that they will be out of stock in week 5. As a result, material requirements planning is projecting that they will have to place an order for 75, the fixed order quantity for the item. The #4651 planned order release line then becomes the projected requirements for the #2156 trigger assembly.

Trigger assemblies #2156 are on order and due in week 4. It's not convenient to stop this order, which is in process at the supplier's plant.

**Figure 8-11
Engineering Change—FPO Technique**

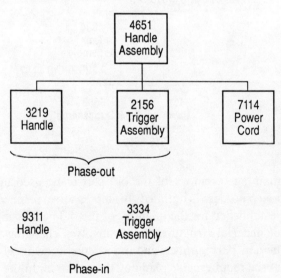

Figure 8-12
FPO Method—Material Planning Pre-FPO

Part No. : __4651__ Lead Time : __1__ Order Qty. : __75__
Handle Assembly

		PERIOD						
		1	2	3	4	5	6	7	8
Projected Gross Requirements			10	20	35	15	40	25	10	55
Scheduled Receipts			75							
Projected Available Balance	5	70	50	15	0	-40 / 35	10	0	-55 / 20
Planned Order Release						75			75	

Part No. : __2156__ Lead Time : __6__ Order Qty. : __100__
Trigger Assembly

		1	2	3	4	5	6	7	8
Projected Gross Requirements					75			75	
Scheduled Receipts					100				
Projected Available Balance	10	10	10	10	35	35	35	-40 / 60	60
Planned Order Release		*100							

*Exception Notice — Place Order

Thus, they will have 110 units of #2156 in week 4 to meet their projected need for 75. The CCS planner then decides to modify the MRP process. Instead of building 75, they will build 110 #4651 handle assemblies with the old #2156 trigger assembly. To do that, the planner uses a firm planned order.

As Figure 8-13 shows, the planner establishes a scheduled receipt due to be received in week 5 for a quantity of 110 handle assemblies. It is coded with an *F*, for firm planned order. Inside the CCS software, this order is treated like any other scheduled receipt, except that it has a special status. It is simply a quantity that CCS plans to build and a date

Figure 8-13
FPO Method—Material Planning Modified with FPO

Part No. : **4651**   Lead Time : **1**   Order Qty. : **75**
Handle Assembly

		PERIOD						
		1	2	3	4	5	6	7	8
Projected Gross Requirements		10	20	35	15	40	25	10	55
Scheduled Receipts		75				110 F			
Projected Available Balance	5	70	50	15	0	70	45	35	-20 / 50
Planned Order Release								75	

Part No. : **2156**   Lead Time : **6**   Order Qty. : **100**
Trigger Assembly

		1	2	3	4	5	6	7	8
Projected Gross Requirements					110				
Scheduled Receipts					100				
Projected Available Balance	10	10	10	10	0	0	0	0	0
Planned Order Release									

Part No. : **3334**   Lead Time : **6**   Order Qty. : **100**
Trigger Assembly

		1	2	3	4	5	6	7	8
Projected Gross Requirements								75	
Scheduled Receipts									
Projected Available Balance	0	0	0	0	0	0	0	-75 / 25	25
Planned Order Release		*100							

* Exception Notice — Place Order

Managing Engineering Change Control 181

on which the company plans to build it. The firm planned order is not released to the factory. It does not generate paperwork or pick lists.

By using the firm planned order, CCS effectively overrides material requirements planning's normal planning logic. The reasons for using the firm planned order have nothing to do with material or capacity requirements. They have to do solely with an engineering change. The objective is to set up a requirement for 110 #2156 trigger assemblies in order to exhaust their stock.

The mechanics of this process involve the requirements file (see Figure 8-14). The firm planned order establishes a series of time-phased requirements in the requirements file, using the current bill of material. The next step is to update the bill of material to the new components. Once the material requirements planning gross-to-net process runs, two things happen: (1) Time-phased requirements come in from the requirements file, so that CCS is planning the requirements for 110 #2156 trigger assemblies; and (2) future plans beyond the firm planned order explode through the current bill of material.

Figure 8-14
FPO Method—The Requirements File

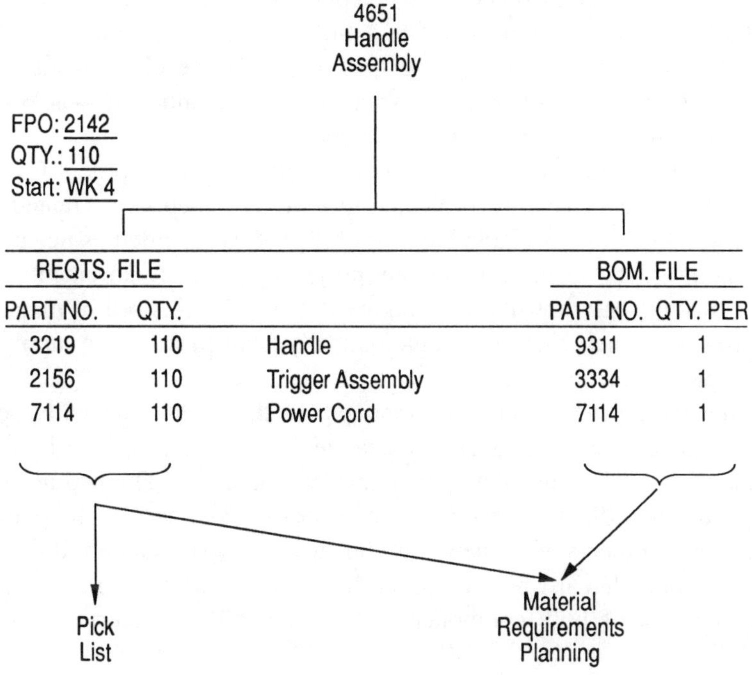

4651
Handle
Assembly

FPO: 2142
QTY.: 110
Start: WK 4

REQTS. FILE			BOM. FILE	
PART NO.	QTY.		PART NO.	QTY. PER
3219	110	Handle	9311	1
2156	110	Trigger Assembly	3334	1
7114	110	Power Cord	7114	1

Pick List

Material Requirements Planning

Of course, when it is time to release an actual work order to the factory for the 110 #4651 handle assemblies, CCS staff must create their pick list from the requirements file. They can't explode the bill of material file, since it no longer contains the configuration they wish to build.

If you only have enough software money to spend on one method of engineering change control, you might consider the firm planned order. You can do more with it, in more circumstances, than with any other tool. The firm planned order is, however, a people-intensive tool. For example, were schedules to change, all firm planned orders would have to be changed manually. It's gratifying to put an effectivity date into a product record and have the computer plan requirements accordingly. But in sheer capability, the firm planned order is a much more powerful tool that can accommodate a wide variety of changes.

Use-Up Technique

Since material requirements planning first came into use, people have used a tool known as the *use-up* to handle engineering change control. To illustrate this technique, let's consider the engineering change in which the #2178 plastic housing is replacing the #1196 metal housing in CCS's parent assembly #3804. A bill of material is established that structures the new housing as a component of the old housing (see Figure 8-15). The lot size for #1196—the old housing—is set at discrete, and the lead time is set at zero.

As material requirements planning develops planned orders for the #3804 housing assembly, projected gross requirements are created for #1196. When the available balance of #1196 is recorded as negative, planned orders result that create projected gross requirements for #2178—the new housing (see Figure 8-16). With this bill of material approach, the new material is automatically planned when the supply of the replaced part is *used up*.

The good news about this approach is that it automatically self-adjusts as demands or inventory change over time. You don't have to manage dates, because the dates manage themselves. The bad news is that whenever the new part does not replace the old part in all applications, requirements of the new component will be overplanned. It should also be noted that any product cost roll-ups done with this structure will add the cost of both components. Also, the bill of material is truly incorrect and could mislead anyone inquiring into this particular configuration. Getting the bill of material deleted after the new part is phased in is also a problem for some companies.

HISTORICAL RECORDS

An area of growing concern to companies is the ability to maintain accurate historical records about the material content and process performance of their products. The detail required in these records will vary according to a company's requirements. Maintaining such information is a routine activity in food and drug manufacturing companies, in those companies involved in aerospace and defense contracting, and in several other manufacturing operations. To automate this historical information, companies must make sure that the necessary data are recorded as the product moves through its manufacturing cycle.

There is a cost trade-off here. A company must assess the business risk of not having the detailed level of documentation versus the cost of

**Figure 8-15
Engineering Change with Bill of Material Structuring Technique**

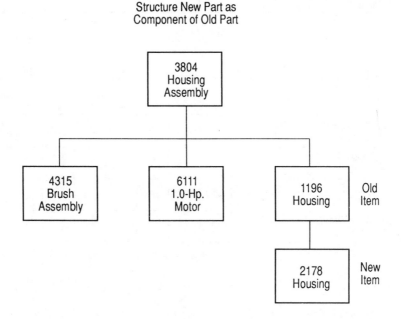

Figure 8-16
ECN Material Planning with Bill of Material Structuring Technique

3804 Housing Assembly	PERIOD							
	1	2	3	4	5	6	7	8
Planned Order Release	20	30	15	25	20	30	10	40

Part No.: 1196   Lead Time: 0 ~~2~~   Order Qty.: LF Lot ~~80~~

Housing

	PERIOD							
	1	2	3	4	5	6	7	8
Projected Gross Requirements		20	30	15	25	20	30	10	40
Scheduled Receipts								
Projected Available Balance	90	70	40	25	0	-20 / 0	-30 / 0	-10 / 0	-40 / 0
Planned Order Release						20	30	10	40

Part No.: 2178   Lead Time: 4   Order Qty.: 100

Housing

	1	2	3	4	5	6	7	8
Projected Gross Requirements					20	30	10	40
Scheduled Receipts								
Projected Available Balance	0	0	0	0	0	-20 / 80	50	40	0
Planned Order Release		*100						

*Exception Notice—Place Order

necessary record keeping. Product liability concerns frequently dictate the need for detailed records. In many companies, the decision is controlled by regulation or customer requirement.

MAKING IT HAPPEN

People have to make decisions and establish rules on how to investigate, evaluate, and ultimately manage engineering change. Making decisions isn't enough without having the proper policies and procedures to implement these changes in a timely manner. This is why the engineering change coordinator and engineering change board are so important. The process also requires performance goals and measurement of success. People and process are 90 percent of an effective engineering change system. With correct documentation and data, the remaining 10 percent—the software techniques that assist in executing the material change—is relatively easy to implement.

Making it happen is all about the people in the company deciding that engineering change control is important and that it requires a formal system to make it work. Once that commitment is made, the right organization and good performance measurements need to be installed to ensure that engineering change control is successful and that the integrity of the company's data foundations are maintained.

Part of making change in an engineering change control process, or in any process involving the way people operate, is gaining ownership and support for the new way of doing business. The next chapter outlines a proven way of successfully implementing change in a manufacturing company.

Chapter 9

Implementing Change

The readers of this book will have various needs. Some are looking for a comprehensive understanding of manufacturing data foundations. Others may be looking for insights into specific approaches, such as structuring bills for by-product situations or handling pseudos. Still others may be implementing major systems such as MRP II.

If you are implementing MRP II, you will want to pay special attention to the last section of this chapter. But all the sections that follow should prove helpful for any change that you might be considering.

PEOPLE: THE MAIN INGREDIENT FOR CHANGE

Implementing change in a company's data foundation, or in the processes used to create or maintain that foundation, can be difficult. Regardless of the scope of the change—adding data to support a new subsystem capability, implementing major systems such as MRP II, or simply getting the data base correct and complete for existing systems—the task can be a challenge. There are always issues concerning the fit of company information to the new data file formats. Some of the issues involve the people whose job it is to support data requirements, get the data right, and maintain data integrity after the system is implemented.

Although implementing change may not be easy, it is certainly not impossible. Following a proven methodology for installing new systems greatly increases the chance for success. This chapter will explore a step-by-step proven process to implement change.

In implementing change, the main ingredient is people. The key people will be the decision makers or department heads who will be most directly affected by that change. These managers, or key personnel representing these managers, form a task team to define the specific changes required, the benefits and impacts of the change, and the major implementation steps that will be required to achieve the change. It is most important that every affected organization be represented on the task team. One person might represent more than one department, but every department must take part in designing the change. If not, that department will be very reluctant to accept and support the change. Figure 9-1 illustrates the likely makeup of data foundation task teams formed to implement preventive maintenance planning or new shop floor systems or to restructure the bills of material.

Once the proper task team is formed, the next step is team education. Typically, the team leader and a couple of other key people attend an outside class on data foundations to broaden their understanding of the subject and to gain insight into the approaches taken by other companies to the particular problem at hand. This group will then educate the balance of the team and the affected department heads. Video education materials are often used. The objective is to broaden thinking and facilitate new ideas on how traditional company approaches might be replaced with improved operating practices.

Once educated, the team needs to review what they have learned and

Figure 9-1
Typical Data Foundation Task Teams

Task Team Makeup

	PREVENTIVE MAINTENANCE PLANNING	SHOP FLOOR SYSTEM	RESTRUCTURE BILLS OF MATERIAL
Design Engineering			X
Manufacturing Engineering	X	X	X
Sales & Marketing			X
Sales Order Processing			X
Planning & Scheduling	X	X	X
Production Operations	X	X	X
Finance		X	X
Quality Assurance		X	X
Service			X
Information Systems	X	X	X

All Affected Departments Represented

match it against current operating practice, identify what works well and what doesn't, and develop proposals on new ways of operating. This requires brainstorming, investigation, and often good constructive debate.

The task may be easy, as in simply identifying new data requirements and assigning responsibility; or the task may be very complex, as in gaining control of an out-of-control engineering change system. Regardless, the process is the same. A series of business meetings are held to discuss the changes required and to gain consensus of the department heads that the proposed changes make sense, are understood, are accepted, and will be supported by the organization (see Figure 9-2).

**Figure 9-2
Series of Business Meetings**

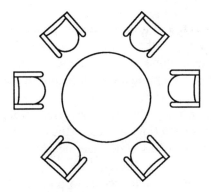

All Key Users Represented
Brainstorm/Investigate/Debate
Propose Change

- Add Value
- Eliminate Waste

Gain Support, Commitment, and Enthusiasm

- Middle Management
- Senior Management

Management Consensus, Ownership
Priority, and Game Plan

- System Definition
- Benefits/Impacts
- Implementation Plan

As change is proposed, the benefits and impacts need to be identified. All change must be tested to ensure that value is added and waste eliminated. If the proposed changes do not demonstrate good benefit, the task should be reevaluated and possibly abandoned. If benefits are projected, the requirements to implement the changes should be identified and reviewed with responsible department heads to ensure that the requirements are supported with resource availability and commitment.

To make this process work, team and department heads must communicate frequently. Another key element to the success of this process is department head resolution of any problems that cannot be resolved by team members. One way to improve team/department head communication and problem resolution is to appoint the affected managers to the task team.

INTERNAL EDUCATION

Once benefits, impact, and implementation plans indicate a clear direction, it is time to share the change proposals with the department organizations. This is done as part of the inside education process, which incorporates communication, learning, discussion, feedback, and action.

Internal education is conducted as a series of department business meetings. Since these meetings are led by the department manager, they might also be thought of as staff meetings concerning the new data foundation requirements (see Figure 9-3).

The first step of inside education is to present some of the broader ideas to which the managers were exposed in their initial outside education programs. The objective is the same—broaden knowledge and facilitate an education environment where change can be explored and accepted. Again, video, when available, is an excellent and consistent medium to present the concepts and principles that need to be understood before specific change is introduced. Well-selected readings are also valuable.

The next step of the internal education process is the discussion of the new systems to be implemented. Note the use of the word *discussion*—not *presentation*. The department manager is prepared to lead this discussion, since this is the person who participated in the development of the new system, clearly understands and supports the specific changes that will impact the staff, and understands the benefits to be

**Figure 9-3
Broad-based Education**

- People Are the "A" Team
- Series of Meetings, People Who Work Together
- Led by Management
- Discussion, Not Presentation
- Feedback Encouraged and Acted Upon
- Internal Education
    - After Task Team Design
    - Prior to Implementation

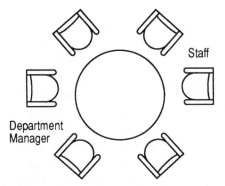

achieved. The manager is also prepared to discuss the implementation responsibilities of the organization.

The most significant element of these discussion sessions is feedback. Feedback comes fairly easy within the staff itself, since it is part of the daily manager/staff communication activity. Three important types of feedback are sought. The first is input that there is something wrong with the changes being proposed or that there could be improvement in the changes being proposed. This is the quality process in action. The staff is being asked to "bulletproof" the new system—to take some shots—to share with management what won't work and what could work better. Since internal education comes immediately after task team and department head project definition, the timing supports revisiting the design and incorporating the good ideas of the world's leading experts—the users themselves.

When the inside education process generates good ideas, it is important that these ideas are acted on. With feedback comes the department manager's responsibility to react and respond to the staff in a timely manner. If staff ideas are ignored, if the response is slow, or if most of the recommendations are rejected, change acceptance will suffer. If such a lack of support to staff feedback is the norm, then a successful implementation and the benefits achieved will probably be less than expected.

A second important feedback is input that one or more persons are uncomfortable with the changes proposed. There is nothing wrong with the new system design, nor are there any better ideas; the individual simply does not feel good about the change. This feedback often occurs when someone's job is changing or a department is taking on new responsibility or transferring responsibility. In this situation, the changes will probably be implemented; but through the internal education process, behavioral bottlenecks to change will be identified. Being aware of them, the manager can, through more education, more discussion, or simply more attention, try to reduce the concerns and improve the feelings about supporting the new system. But this only becomes possible when management is aware that these feelings exist.

The third type of feedback is acceptance. This will be easy to detect as the staff begins to show enthusiasm and demonstrates an eagerness to get started. Enthusiasm, acceptance, and support result when the learning environment is one where people know each other; where questions are raised and answered; where benefits are reviewed so that people not only understand what is going to change, but also why; where department and personnel impacts are presented for open discussion; and where people are encouraged to present ideas and those ideas are implemented.

The internal education process ends with the initiation of the implementation process. The right time to start implementation is when understanding and acceptance are achieved—not at some later date. Typical implementation steps may include software training, data collection, data accuracy improvement, and new procedure development.

The final step is the actual implementation of the new data foundation capabilities. Implementing change is never easy. Using enlightened people as task teams in the manner described in this chapter, however, will lighten the task and improve the chances for success.

IMPLEMENTING THE DATA FOUNDATION FOR MANUFACTURING RESOURCE PLANNING AND JUST-IN-TIME

Fortunately, there already exists a proven path to implement successfully an integrated MRP II system. For our purposes, we will not explain in depth all the steps necessary for implementing MRP II, but will concentrate primarily on how to implement the data foundation as part of the process.

It is important, however, to gain a perspective of the overall implementation process shown in Figure 9-4. Audit assessment I identifies the need and is followed by first-cut education, which primarily involves top management and a number of people from middle management. This provides the understanding of what the tools of MRP II and JIT/TQC can provide to meet a company's needs. This is also when these individuals come to terms with the costs, the paybacks, and the importance of top management leadership.

The next implementation steps produce the vision statement and a cost/benefit analysis. Although much of the detailed work for the cost/benefit analysis is done by the middle management team, top management must also go through the process, agree to the numbers, and commit to the implementation.

It is then up to the top management team to set up a project organization to implement MRP II. This involves selecting a full-time project leader and creating a steering committee to guide and direct the MRP II implementation from a senior management level.

The steering committee is made up of the top management group and the project leader. It is chaired by the president, managing director, or general manager. The project leader then establishes a cross-functional project team. Spin-off task forces or groups are also formed to link into the project team (see Figure 9-5). Typical spin-off task forces might include any of the following:

- Master production scheduling
- Data foundations—bills of material, routings (one or more teams)
- Inventory record accuracy
- Purchasing
- Manufacturing
- Financial planning

Although this discussion is focused on the activities of the data foundations spin-off task force, some interrelationships will develop

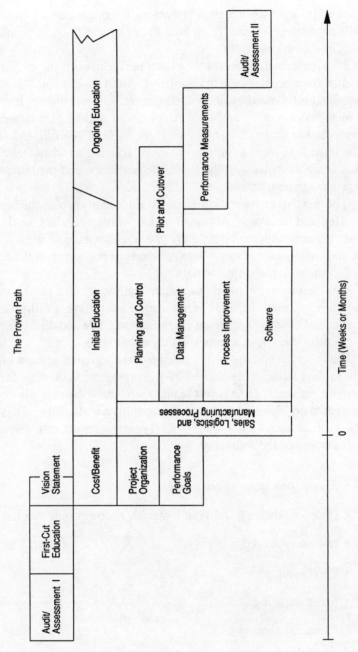

**Figure 9-4
The Proven Path**

The Proven Path

Figure 9-5
Project Organization

Company-Wide MRP II

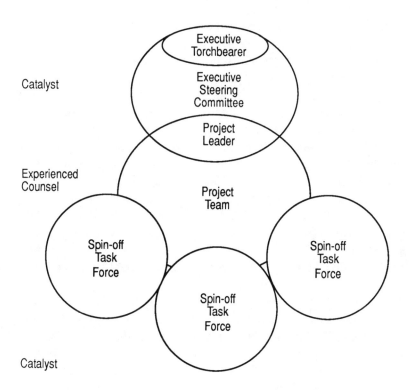

with other task forces. For instance, structuring the bills of material goes hand in hand with structuring routings. Issues like scrap, yield, and shrinkage must also be considered. The data foundations group will also need to work closely with the manufacturing task group.

The master production scheduling task force will consider whether the company's products are make-to-stock, make-to-order, finish- or assemble-to-order, or custom products.

In a finish-to-order environment, planning bills of material may need to be used, and they will have a major impact on structuring the bills of material. New or custom products using the planning bill of activities discussed in Chapter 6 will impact the bill of material task force.

In essence, the data foundation task group is a team that will work primarily on its own, but it cannot work in isolation. Typically, the task force members are part-time and come from the user network. The responsibility for interlinking the various task groups is with the project team members.

Once the members of the task force are educated, their primary task is to evaluate current company practices, determine action items and devise an implementation plan. Figure 9-6 illustrates the general data foundations implementation framework. Three primary issues—accuracy, completeness, and structure—need to be addressed in regard to the bill of material and the company's data foundation. In very large companies, two task force teams might be formed to work on these issues: one to examine structure and the other to assess accuracy and completeness. Medium-sized companies might only need one task force team to perform all three activities. Some small companies combine the data foundation task force and manufacturing task force into one.

Even though, in most cases, the data foundation task teams are led by design or engineering, the responsibility for this process does not rest solely on their shoulders. Top management is still responsible for leading the drive toward an effective implementation and establishing one company bill of material data base. Sales and marketing will have to be involved with sales configurations and product options, and their input regarding any restructuring of the data foundations and bills of material will be needed. Planning will also be impacted, especially when we look at the way we sell our products; their input will be required in areas such as defining the planning bill of material. Manufacturing also has a say in what happens, since it pertains to production schedules, shop floor control, material supplies, and the modularization process. Finance is concerned with accuracy and how manufacturing costs are impacted by what goes on the bills of material. Service parts must also be included in defining the way we sell parts or spares, as well as in defining configuration control. Data processing management information systems will be involved in the total integration of the single-company data foundation.

Obviously, implementation is a team effort.

PROJECT ORGANIZATION

Data foundation users need to be represented on the task team, but size is an issue. If the team becomes too big, one result can be more talk than action. When necessary, it's better to identify a core group, with addi-

Figure 9-6
Data Foundation Task Force Plan

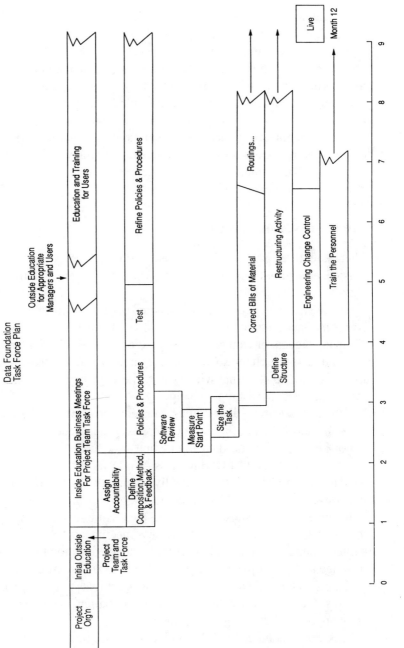

tional functions represented on specific issue as required. The core group should be composed of people from the design/technical department, manufacturing, planning or material control, data processing, and finance, and there should also be a project team member responsible for coordinating tasks that spill over into other task forces.

Other departments should be represented on the team on an ad hoc basis when their interest is being addressed. For example, production or industrial engineering needs representation on structuring issues, and quality control needs to provide input on accuracy, yield, scrap, and shrinkage. Sales, marketing, and order entry need to be involved on structuring to support the service process.

Education and Training

There are three stages of initial education before going live: initial outside education, inside education, and education and training for users.

The task force needs initial outside education on structuring the company data foundation to support MRP II and JIT/TQC. Within the company, agents of change need to be created who can challenge tradition and facilitate change. Companies that fail to do this risk implementing a data foundation that will support the way they do business today rather than accomplishing a change.

Outside education also equips the team to develop their detailed implementation plan. Figure 9-6 depicts generalized milestones and major activities for the data foundation task force in a company that decides to implement MRP II to class A in 18 months. Certain activities need to be completed before going live in month 12. The task force's job is to develop a plan for its own company.

Inside education is also needed. To augment understanding, the task force needs to create a series of business meetings (like those described earlier), using a vehicle such as tailored video education to reinforce the message and develop detailed knowledge of its own company issues and problems. This education and training culminates in a test between months 4 and 5 to see how the data foundation integrates into one company way of running the business.

Task force members and the project team are now equipped to teach managers new ways, and they, in turn, can teach their people, the users.

Some functional experts—engineering change coordinators, bill of material administrators, master schedulers, configuration control specialists, and data base administrators—will need outside education as well as inside education. Other bill of material users will require inside education and training from their managers (who have already been educated by project team and task force leaders) on the new way of working.

Ongoing education after implementation is essential. People come and go. Outside and inside education is essential for new personnel. A practical way of working is to tie education and training needs into job descriptions.

Assigning Accountability

The task force must assign accountability for

- bill of material accuracy
- routing and work center accuracy
- engineering change coordination
- policy and procedure

Accountability is assigned not only to task force members but to user managers as well. The director or vice president responsible for these areas must also accept accountability for leading the new way of managing with his or her group.

Defining Composition, Method, and Feedback

The task team needs to determine what the data foundation requirements are, how to audit the integrity of those data foundations once they have been implemented, and how to manage error identification and correction. These areas were discussed in detail in Chapters 2 through 4. The task team's responsibilities are covered here.

These tasks can be highly charged and emotional. Debate can go on and on with no definitive conclusion. Therefore, it's recommended that a limited period of four weeks be allowed for their completion.

The first issue that a task team needs to identify is the company's data foundation requirements—its *composition*. What needs to be included on the bills of material? What types of bill of material formats apply? How should the bill of material be structured? Routing, item master, and drawing system requirements also need to be defined.

Method refers to the audit process chosen to monitor the integrity of the MRP II data foundation. It involves the auditing of the bills of material, routings, item master and work center master files. The task team responsibilities include defining the audit methods and the reporting requirements and determining the audit responsibility.

An overriding principle when operating MRP II is "Silence is approval." Unless there is *feedback*, people will assume that the data foundations are complete and accurate. There is a need to *speak up* or initiate corrective action whenever the software identifies an error or edit condition. Someone must also say something when audits identify data integrity issues or when it is determined that needed information is missing or erroneous information exists in the data foundations.

To support data foundation error resolution, the task team needs to identify the needed communication, correction, and feedback policies and procedures. These will then be refined and simplified by the managers and the users through the initial education and training programs mentioned earlier.

Software Review

The task force must lead the software review to understand how the software operates and how it serves the data foundation requirements defined in the *composition* phase of the project.

Following the actual testing of the software, which occurs during the *test* phase of the project, a decision has to be reached on modifications or enhancements. The task team needs to be sure that the correct software data files are available and that they meet the users' needs. The team must also ensure that the users are properly trained in software use and maintenance requirements before going *live*.

Measuring a Start Point

This activity is to put a "stake in the ground" and size the task of cleaning up or establishing the required data foundation files. To do this, existing bills of material, routing, and item master files need to be audited against the new MRP II system data foundation needs.

Typically, a sample set is selected (often 100 bills of material and their associated routings and item master records), and it is audited to determine the scope of the effort.

SIZING THE TASK

Once a starting point is established, as well as the amount of existing data file work that needs to be accomplished, the resources required to clean up, modify, or add the new data requirements can be identified. A detailed, time-phased plan is then developed to complete the required data foundation preparation.

The plan must be expressed in measurable amounts of work or in meaningful milestones. For instance, to say that it is necessary to correct 15,000 bills of material in seven months is not useful. A measurable quantity would be 500 bills of material corrected and audited every week for the next seven months. This work load can be monitored easily. Progress should be reported weekly to the project team, and sample audits of corrected data should be conducted regularly with results reported.

DEFINING STRUCTURING ACTIVITY

Structure, or architecture, of the bills of material must be reviewed in conjunction with other task forces.

The master scheduling task force needs to determine which products are to be finish-to-order and if planning bills of material that support forecasting, order entry, and finished schedule paperwork (pick lists, finishing routings, and configuration control) will be needed. Use of pseudos, phantoms, and bill flattening must be consistent.

The financial task force needs to be involved to ensure that cost rollup is supported.

The manufacturing task force needs to discuss with the task team the integration of structured bills of material and routings. Flattening bills of material and simple routings are important in cellular manufacturing, continuous batch processes, and flow processes.

Any necessary restructuring tasks must be defined and resources allocated once the task is sized.

ENGINEERING CHANGE CONTROL

Policies and procedures must be developed to describe how engineering changes will be controlled. Here are some of the issues to be considered:

- Organization of engineering change request
- Procedures outlining investigation, appraisal, justification, authority levels, decision making, and priorities
- Impact on planning
  —Activities, timing, and responsibilities
  —The most appropriate techniques—effectivity date or lot, or firm planned order method
- Execution of engineering change, including measurement of progress and costs
- Postimplementation audits and the all-important question: Did the change accomplish what it was set up to do?

Appropriate documentation must be developed and the system technique reviewed in the test pilots before implementation.

The organization needs to agree, and the responsibilities and the authority of the engineering change coordinator and bill of material administrator, if any, must be in place before the conference pilot is used.

Making It Happen: The ABCD Checklist

The Oliver Wight ABCD Checklist provides a well-accepted measurement of a company's progress toward class A planning and control processes and continuous improvement. The checklist contains a series of key element measurements (overview questions) with hundreds of supporting audit points (audit questions).

Key element measures involving data foundation management include the following:

- A single set of numbers is used by all functions, with the operating system providing the source data used for financial planning, reporting, and measurement.
- All phases of new product development are integrated with the planning and control system.
- Where applicable, engineering activities in support of a customer order are integrated with the planning and control system.

- The planning and control process is supported by a properly structured, integrated set of bill of material, routing, and related data.
- Data integrity is measured against preestablished tolerances and meets accuracy requirements, including the following fundamental standards: bills of material, formulas, recipes, etc., 98 to 100 percent; routings, 95 to 100 percent.
- There is an effective process for evaluating, planning, and controlling changes to existing products.

Each of us is challenged to test our company's practices against these benchmarks for success. Getting the data foundation right and keeping it right are important elements. But even more important is that a good data foundation makes it possible to achieve other elements of class A success.

THE SOLID MANUFACTURING DATA FOUNDATION

The solid manufacturing data foundation must correctly define all the company's products and processes—bills of material, routings, item information, work center information, engineering change, and drawing systems. It must be organized to facilitate all users—forecasting, master scheduling, configuration control, order entry, material planning and acquisition, production scheduling and control, product costing and manufacturing, and finance and service parts. It must be one set of accurate company numbers. It must include a mechanism for effective change control. In short, it must be a solid manufacturing data foundation—a robust recipe indeed.

Appendix A
By-products and Co-products

Many manufacturing processes result in the production of by-products and co-products. For example, very little leaves a beef slaughterhouse as waste. Although beef is the prime product, many by-products result.

Semiconductor manufacture often results in by-product or co-product items in addition to the prime product. Testing, sorting, and grading untested devices may yield products suitable for high-reliability military applications as well as products suitable only for less demanding commercial or industrial uses. Toxic and hazardous wastes are also by-products of various manufacturing processes. It's becoming increasingly important in many industries to manage this type of by-product output.

Figure A-1 shows a common convention for displaying the bill of material for product A with four component ingredients. Using a chemical process example, let's look at what happens when the product A is not the only product resulting from the process. The reaction involves ingredients 1, 2, 3, and 4 and produces 60 percent prime product, 30 percent co-product X, and 10 percent by-product Y (see Figure A-2).

The logic involved in structuring this bill of material and using it for planning purposes is very similar to the disassembly bill of material logic discussed later in Appendix C.

206   By-products and Co-products

Suppose that the ingredients for product A are mixed in the following ratios:

Quantity	Ingredient/Component
30 kilograms	1
60 kilograms	2
60 kilograms	3
60 kilograms	4

The chemical reaction yields the following results: 90 kilograms of A, 45 Kilograms of X, 15 kilograms of Y, and some unplanned waste. The bill of material is then structured (see Figure A-3). The quantities per relate to the parent/component relationships needed to produce one unit of prime product A. Ingredients 1, 2, 3, and 4 are coded as components, while X is coded as a co-product and Y as a by-product in the bill of material record.

Figure A-1
One Parent from Several Components

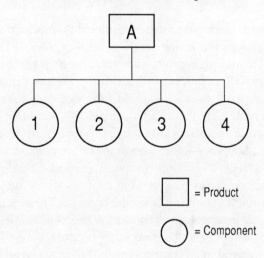

**Figure A-2
By- and Co-product Items**

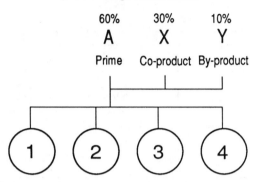

**Figure A-3
By- and Co-product Bill of Material Relationships**

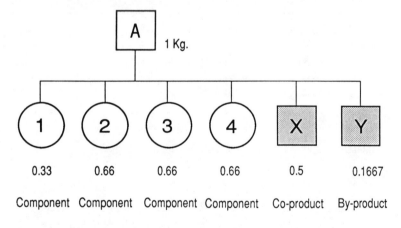

Figure A-4
Material Requirements Planning for Standard Components

Part Number.: **A**   Lt: **1** Week   Order Qty.: **90**   On Hand: **100**

		1	2	3	4	5	6
Gross Reqts.			50	40	50	30	60	60
Sched. Receipts							
Proj. Avail. Bal.	100		50	10	90 / -40	20	90 / -40	90 / -10
Planned Orders				90		90	90

Qty. per on BOM. = 0.66

Part Number.: **2**   Lt: **4** Weeks   Order Qty.: **200**   On Hand: **130**

		1	2	3	4	5	6
Gross Reqts.			60		60	60	
Sched. Receipts							
Proj. Avail. Bal.	130		130	70	70	10	200 / -50	150
Planned Orders			*200				

*Message: Open an order for component ②, due Period 5.

Figure A-4 shows how MRP uses the standard parent/component relationship in the bill of material to explode planned orders for prime product A to gross requirements for ingredient 2 using planned order start dates for A (shaded area in the figure).

There are two significantly different ways to handle planning of by-products and co-products through (1) negative gross requirements and (2) a special type of scheduled receipt.

By-products and Co-products

Negative gross requirements involve by-products and co-products in the recipe, or bill of material, with a negative quantity per. This shows a time-phased buildup of the by-products and co-products in the planning system (see Figure A-5).

The negative gross requirements method is more popular than the special scheduled receipt method, because it doesn't require any modification to typical MRP II software. However, there are three problems with this process:

1. It can be confusing when, for example, gross requirements of 105 for by-product Y (coming from another demand stream in MRP) are combined with -15, and the planner sees a net gross requirement of 90 (see Figure A-6).

2. Negative gross requirement does not account for lead time properly. The negative gross requirement for the by-product Y appears at the start of the order for main product A, rather than on the due date, which is when the by-product will be available (see Figure A-7).

3. Using a negative quantity in the bill of material can be a problem, because it can provide erroneous costing calculations.

Figure A-5
By-product Planning Using Negative Gross Requirements

A
|
Qty. per 0.1667
|
Y

Product A	\multicolumn{6}{c}{WEEKS}
	1	2	3	4	5	6
Gross Reqts.	50	40	50	30	60	60
Sched. Rcpts.						
Proj. Avail.	50	10	50	20	50	80
Planned Ord.		90		90	90	

By-product Y	\multicolumn{6}{c}{WEEKS}
	1	2	3	4	5	6
Gross Reqts.		-15		-15	-15	
Sched. Rcpts.						

By-products and Co-products

Figure A-6
Net Gross Requirement Confusion

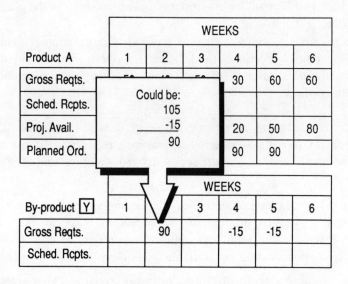

Figure A-7
Lead Time Problem

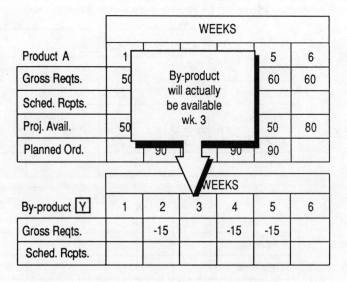

For many companies, the disadvantages in the preceding method are outweighed by the costs of modifying software to provide the scheduled receipts method. The scheduled receipts method is a better choice of approach than the negative gross method, providing a company's software already has this capability.

In the scheduled receipts method, Y is coded as a by-product of A with a normal quantity per 0.1667. The by-product coding enables planned orders for A to explode in scheduled receipts for Y (see Figure A-8).

The scheduled receipts for Y are a special kind of scheduled receipt. They're not released and will remain until the planned orders for A are released. When a planned order for A becomes a scheduled receipt, the scheduled receipt for Y also becomes *released*.

With this method, planners can easily see the difference between demand and supply for the by-product, whereas the negative gross requirement method presents supply as a negative gross requirement, creating potential confusion.

Within the scheduled receipt method, planned orders for A do not explode to by-product Y at the start date. They explode at the due date.

Figure A-8
By-product Planning Using Scheduled Receipts Method

A
| Qty. per 0.1667
Y

Product A	\multicolumn{6}{c}{WEEKS}
	1	2	3	4	5	6
Gross Reqts.	50	40	50	30	60	60
Sched. Rcpts.						
Proj. Avail.	50	10	50	20	50	80
Planned Ord.		90		90	90	

By-product Y	\multicolumn{6}{c}{WEEKS}
	1	2	3	4	5	6
Gross Reqts.						
Sched. Rcpts.			15		15	15

In periods 3, 5, and 6, 90 kilograms of A are planned to be available, and 15 kilograms of Y will be available on the same dates as A. The lead time for making by-product Y is accounted for properly (see Figure A-9).

In modifying software to support the scheduled receipts method, it is usually necessary to review the standard firm planned and work order release logic in the software when dealing with by-products and co-products. As schedules are firmed or released for prime product A, standard logic would establish requirements for ingredients 1, 2, 3, and 4. Special by-product and co-product logic must be in place to establish firm and released scheduled receipts for products X and Y. Once this is done, normal receipt activity would be used to receive by-product and co-product into stock.

In addition, it is usually necessary to review any standard cost accounting processes in use. Normal standard cost roll-up techniques need to be adapted for the existence of by-product and co-product items in bills of material.

There may be situations where the demand for by-product Y exceeds the supply of Y, which will be by-produced with prime product A. When this happens, planners will have to check the plans for both Y and A.

Co-products are handled in much the same way as by-products. The difference is that co-products have a higher yield than by-products, and more emphasis is needed from the planning department. Balancing supply and demand in a co-product situation can be more complex because the ratios of production may not agree with marketplace demands. For this reason, many companies customize their planning exception reporting so that all interdependent items (main product, co-products, and by-products) are visible simultaneously.

A large chemical company produces six batches of main product to three batches of co-product. Sometimes more of the main product is required; other times more of the co-product is needed. Since these products are stored as molten oils in heated storage, a high demand in co-product would result in main product production exceeding the storage capacity. Planning options in this scenario are either to purchase co-product or involve sales in selling off more main product.

A cheese-producing company uses by-product logic to plan cheese off-cuts as a by-product of its packaging process. Because these bits and pieces are an important ingredient in its processed cheese, the company needs to know the amount in off-cuts that will be generated in a time-

Figure A-9
By- and Co-product Planning

Product

Part Number: A   Lt: 1 Week   Order Qty.: 90   On Hand: 100

		1	2	3	4	5	6
Gross Reqts.			50	40	50	30	60	60
Sched. Receipts							
Proj. Avail. Bal.	100	50	10	90 -40	20	90 -40	90 -10
Planned Orders				90		90	90

By-product   Qty. per on BOM = 0.1667

Part Number: Y   Lt: ___ Week   Order Qty.: ___   On Hand: ___

		1	2	3	4	5	6
Gross Reqts.							
Sched. Receipts					15		15	15
Proj. Avail. Bal.							
Planned Orders							

Scheduled Receipt Details

DATE	QTY.	TYPE AND STATUS
Per 3	15	From Prime Product P/N A Not Yet Released
Per 5	15	From Prime Product P/N A Not Yet Released
Per 6	15	From Prime Product P/N A Not Yet Released

214   *By-products and Co-products*

phased manner. Management wants to use these off-cuts before it uses bulk material cheeses.

The bill of material uses a combination of by-product logic and the reprocessing logic discussed in Appendix D for the off-cuts. The off-cuts are treated like phantom items with zero lead times and lot-for-lot or discrete-lot sizing. The bill of material for this procedure is shown in Figure A-10. The company uses the negative gross requirements method. Lead times in packaging are very short, so that lead time distortion is not an issue.

Finally, a low-grade metal-refining company has two distinct processes. Metal scrap is upgraded to a metal-rich ingot, which is then chemically refined to provide a more pure metal. By-products of both processes are low-grade residues that are recycled into the upgrading process. It is essential to plan the time-phased yield for these residues, because they consume a great deal of capacity in the upgrading process. The bills of material for the two processes appear in Figures A-11 and A-12. Because the lead times for both processes are more than a day or two, the scheduled receipts method is used.

Figure A-10

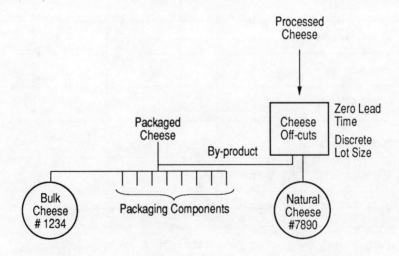

**Figure A-11
"Upgrading" Process Bill of Material**

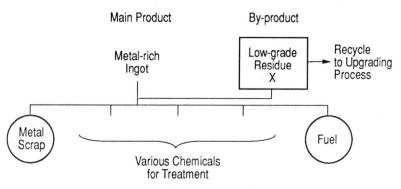

**Figure A-12
Refining Process Bill of Material**

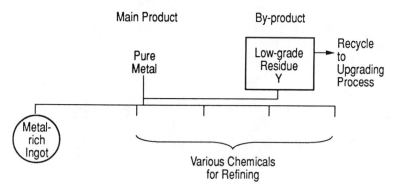

Appendix B

Preventive Maintenance

Preventive maintenance planning and control is an important issue for many manufacturers. Stand-alone preventive maintenance systems are in use in some companies. Other companies, however, utilize bills of material, routings, and MRP II to plan maintenance materials and to schedule the actual maintenance activity.

The planning of maintenance material is usually done by assigning each maintenance activity an item number. A bill of material is created for each preventive maintenance item number that lists all the maintenance materials (oil, belts, indicator lights, etc.) as components to the parent maintenance item (see Figure B-1). The maintenance planner then creates firm planned orders in the material requirements system (independent demand) for the dates the maintenance should be performed.

When material requirements planning is performed, dependent demand planning using the maintenance bills of material will generate time-phased demand for the maintenance materials. Requirements for common materials that support multiple maintenance activities will be consolidated in the process. As these demands are netted against inventories and scheduled receipts, order requirements and reschedules are identified.

Frequently, a given material use is different in each preventive maintenance process. In this case, the quantity per is usually loaded as a percent or decimal quantity to represent the anticipated use of the material. Most companies carry a slight amount of safety stock for percentage error. Maintenance materials with a very low anticipated usage might not be included on the bill of material. They may simply be carried on the item master with a safety stock requirement that will create reorders as the existing inventory is consumed.

Many maintenance materials have a regular usage for unplanned maintenance support. When this happens, the planner may load an independent demand to provide material coverage. The process is similar to entering an independent demand forecast for a service part.

To plan and schedule preventive maintenance tasks, a routing is assigned to the parent item representing the maintenance activity (see Figure B-2). By identifying the parent maintenance item as a *make* part, the capacity planning system will time-phase plan the standard hours of downtime for the appropriate work center. Note that there is only one operation in the routing and that the hours planned are in the *setup* field, not the *run time* field. The hours required are not piece driven. Work center supervisors and production control personnel can now smooth the work center load with both production demand and maintenance downtime visible in detail and summary. When the maintenance order is released, the task with its start and complete dates will also be included on the work center dispatch list if a standard shop floor control system is in use.

Some companies develop custom subsystems in order to report the detailed maintenance schedules and the summarized hours to the pre-

Figure B-1
Preventive Maintenance Bill of Material

Parent Number: M663
Description: P.M. for Large Mixing Machine

COMPONENT	DESCRIPTION	QUANTITY PER
1472	Oil	18.0
1764	Drive Belt	0.5
1961	Sensor A	0.2
1963	Sensor B	0.4
3604	Temp. Indicator	2.0

ventive maintenance planner for capacity planning and dispatch scheduling. This usually requires some programming effort within the shop floor systems. The same information that is posted at the appropriate work center (standard software) needs to be posted to a preventive maintenance file (nonstandard software).

Some companies will create multiple operation steps in their maintenance routings to provide for the capacity planning and scheduling of specific skilled workers, such as pipe fitters, electricians, mechanics, and so on (see Figure B-3). A frequent problem with this is that many systems assume that each routing step is sequential and plan the activities in that manner, when, in reality, the various resources might be working simultaneously. In many applications, this level of error is not a problem. Some software allows operation overlapping, which may accommodate simultaneous multi-operation activity. If simultaneous operations are a problem in maintenance planning and scheduling, and if overlapping is not available, some systems modification will be necessary.

Figure B-2
Preventive Maintenance Routing

Parent Number: M663
Description: P.M. for Large Mixing Machine

OP.	DEPT.	W.C.	DESCRIPTION	SETUP	RUN TIME
10	M	16	Quarterly Preventive Maintenance	42	0.0

Figure B-3
Multiple Operation Routing

Parent Number: M663
Description: P.M. for Large Mixing Machine

OP.	DEPT.	W.C.	DESCRIPTION	SETUP	RUN TIME
10	M	16	P.M. Pipe Fitting	12	0.0
20	M	16	P.M. Electrical	14	0.0
30	M	16	P.M. Mechanical	16	0.0

Appendix C

Refurbishing, Remanufacturing, and Reconditioning

Remanufacturing is the process of taking an existing used item or product and reworking it to a like-new state or to a modified state. This part or product is then equal in its capability to the original equipment or offers like-new capability in its modified form. It may not be new, but in most cases, the warranties that accompany these parts or products are as good as those attached to a new product.

Remanufacturing typically occurs in companies that lease equipment (computers, copiers, etc.), refurbish their own inventories (NASA), resell equipment (office equipment), or are in aftermarket parts sales (automotive). Typically, the cost of remanufacturing is much less than new-build, which provides for attractive product pricing and minimizes the expense of maintaining leased equipment. In some cases, however, heavy remanufacturing is required simply to keep older products in an operational state. Both military equipment and commercial aircraft fall into this category.

In most remanufacturing operations, a company will receive one unit, tear it down, and refurbish the unit to a remanufactured state. In some cases, more items or products enter the process than leave the process because of quality review and scrap. The amount of usable reclaimed material and the costs of the remanufacturing processes need to be watched closely to ensure that the remanufacturing strategy remains a viable alternative to a new-build approach.

The data foundations necessary to support remanufacturing planning and control require some estimating and are rarely precise. Parts yield and replacement rates vary. Processes vary, depending on the condition of the returned product. The actual cost of each product shipped will vary with the amount of new components required or the amount of labor expended in the remanufacturing effort. Data foundations do, however, need to be determined and provide acceptable planning and control averages.

There is another point of caution. Although the finished part or product needs to meet shippable specifications, remanufacturing must remain highly cost effective. Product price is often the basis of competition. For this reason, it is advisable for companies involved in remanufacturing to take a close look at the principles and concepts of Just-in-Time, working at the continual elimination of waste in their processes. With more simplified processes, the cost of maintaining large, complex data files is also eliminated.

ITEM OR PART NUMBERS

Refurbished items may not be given the same part numbers as new parts, since they carry different costs, require different materials, and utilize different processes. Yet, in many applications, once a new or refurbished item is used in reassembly, it loses its unique identity. Some remanufacturers assign a unique, nonsignificant part number to each refurbished item and use item descriptions for part identification. Another popular approach is to use an item number suffix. In this application, a new #123456 might become a #123456-1 in its preteardown state, then a #123456-2 in its refurbished state. Others make this distinction by adding a three-letter descriptive suffix—TDN for teardown, RWK for rework, and REC for reconditioned. This suffix is treated as part of the regular item number field. A similar approach is sometimes used in distribution networks to identify the same items in different distribution centers.

CCS REMANUFACTURING OPERATION

To better understand how this remanufacturing process works, let's use as an example the CCS #4215 brush assembly, which will simply be labeled item A. With high use, the bristles of the brush unit wear out, and occasionally, the brush tube itself becomes damaged. As a result,

CCS has been able to build a sizable remanufactured parts replacement business (see Figure C-1). A remanufactured brush assembly is much cheaper for their customers.

Once the CCS product engineering and planning staffs developed their estimated replacement rates and the associated remanufacturing processes, CCS was able to create the data foundations necessary to support the planning and control requirements for this business.

The first process step was the teardown of the returned units or cores. To encourage core returns, the packaging engineers created a remanufactured parts carton complete with a return label and its own customer identification number. This allowed customers to remove the replacement brush assembly, insert the worn or broken assembly into the same box, and mail it back to the remanufacturing center. For doing this, the customer received a small reimbursement for each core that was returned.

To define the anticipated yield of components from teardown, CCS created the disassembly bill of material (shown in Figure C-2). Part A has been given a suffix TDN to indicate that it is a core in a preteardown state. This allows CCS to give it a separate cost or value and to maintain a separate inventory for teardown planning. Because of special coding in the disassembly bill of material, the company's planning systems (MRP)

**Figure C-1
Assembly Drawing and Bill of Material**

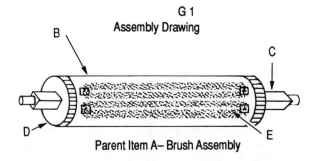

G 1
Assembly Drawing

Parent Item A– Brush Assembly

COMPONENT	DESCRIPTION	QUANTITY PER
B	Brush Tube Assembly	1
C	Brush Shaft	1
D	Brush Tube Endpiece	2
E	Brush Bristles	6

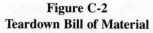

Figure C-2
Teardown Bill of Material

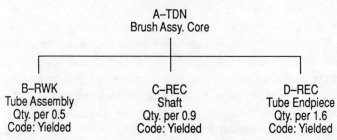

realize that the specified components will be yielded in the manufacturing process and not consumed. In the case of A-TDN, engineering expects to yield:

—B tube assemblies 50 percent of the time. These tubes will require rework, so they are labeled RWK. This allows CCS to assign a cost and control their inventory.

—C shafts 90 percent of the time. Once these shafts have been cleaned, they go to stores for reassembly. By adding REC, it is recognized that these are reconditioned and ready for reassembly. The suffix again allows for stocking and costing.

—D endpieces 80 percent of the time. All they have to do to these endpieces is clean them and they are ready for reassembly (REC).

The manufacturing engineering group also created a routing for the teardown process (see Figure C-3).

Figure C-3
Teardown Routing

OP.	W.C.	DESCRIPTION	SETUP	RUN TIME (PER 100)
10	TDN	Loosen shaft, remove, place in clean basket.	0	0.30
20	TDN	Remove ends, inspect for shaft size, discard rejects, place good ends in clean basket.	0	0.40
30	TDN	Remove bristle assemblies from tubes. Send tube assemblies to stores (RWK).	0	1.10
40	CLN	Wash shafts and ends. Forward to stores (REC).	1.5	0

With these routings, CCS can plan manufacturing capacities, cost work-in-process, and apply standard cost accounting control. They can even use basic shop floor control techniques for scheduling. Note that the routing indicates which suffix is to be assigned as the yielded part leaves the process for stocking. Also note that the cleaning operation uses setup and no run time, since the cleaning time is fixed.

The engineering department next identified the requirement to rework or refurbish the brush tube assemblies. At this point, normal bills of material rather than yield bills of material are used (see Figure C-4). In the brush tube bill of material, part X is a purchase part that normally comes on new brush assemblies. Since the company is refurbishing a purchased part, they do need to buy some belt link rings to support their rework activity.

Figure C-4
Rework Bill of Material

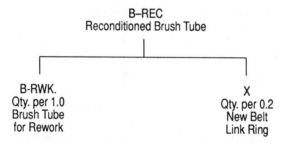

Also, a routing was established to describe the process and provide for detail planning, scheduling, and cost control (see Figure C-5).

At this point, CCS has defined all the material that they expect to

Figure C-5
Rework Bill of Material

OP.	W.C.	DESCRIPTION	SETUP	RUN TIME
10	RWK	Test for roundness and dents; discard rejects.	0	0.20
20	RWK	Remove belt link ring; inspect for damage; discard rejects.	0	0.40
30	RWK	Wash in special cleaning compound.	1.5	0
40	PNT	Paint.	.5	0
50	RWK	Replace belt link ring. Use new if necessary. Send to stores (REC).	0	1.00

Figure C-6
Remanufacture Bill of Material

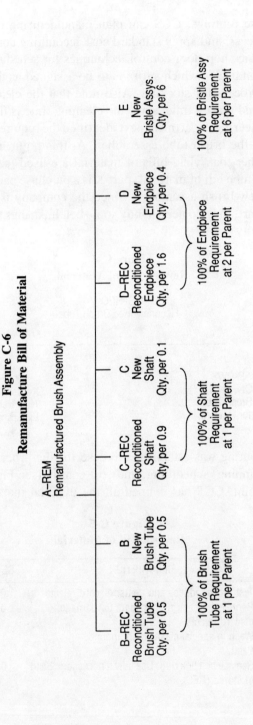

recover from the teardown and reconditioning, but they still require some new parts to bring all their cores to a remanufactured, sellable state. The remanufacturing bill of material indicates the mix of reconditioned and new parts that will be required (see Figure C-6).

Of course, there would also be a routing to plan capacity and schedule reassembly operations. All accounting functions are also satisfied with the bill of material and routing structures, just as they were for teardown and rework. Product costing is also accomplished with normal roll-up routines. The end item cost for the remanufactured brush assembly is displayed in Figure C-7.

PLANNING

With the preceding data foundations in place, CCS can now implement time-phased planning and scheduling using MRP II. The planning process begins at teardown, where the supply of yielded materials is planned. Two specialized software capabilities are required. First, the teardown schedule must post scheduled receipts as opposed to projected gross requirements for yield components. Second, lead time must be used to determine when the yielded material will be available versus when it is required (see Figure C-8).

The teardown master schedule of 100 a week is extended by the quantity per of each yielded component (see Figure C-2). The lead time to tear down item A-TDN results in a two-week offset of component

Figure C-7
Remanufactured Brush Assembly Product Cost

PART	DESCRIPTION	COMPONENT COST QUANTITY PER	ASSY. COST
E	New Bristle Assys.	$ 2.00 × 6 =	$12.00
D	New Endpiece	2.50 × 0.4 =	1.00
D REC	Reconditioned Endpiece	0.30 × 1.6 =	.48
C	New Shaft	2.00 × 0.1 =	.20
C REC	Reconditioned Shaft	0.30 × 0.9 =	.27
B	New Tube	10.00 × 0.5 =	5.00
B REC	Reconditioned Tube	4.00 × 0.5 =	2.00
		Total Product Cost	$20.95

Figure C-8
Teardown Yield Plan

Teardown Master Schedule:
Item: A-TDN
Inv.: 0
Lt.: 2

	WEEKS						
1	2	3	4	5	6	7	8
100	100	100	100	100	100	100	100

Scheduled Receipt Plan

Item		1	2	3	4	5	6	7	8
Item	B-RWK Qty. per 0.5			50	50	50	50	50	50
Item	C-REC Qty. per 0.9			90	90	90	90	90	90
Item	D-REC Qty. per 1.6			160	160	160	160	160	160

availability. Following the teardown process, the yielded components are sent to stores, and their scheduled receipts are closed out with the actual yield achieved, not the projected or planned yield.

Items C-REC and D-REC are now available for use in the reassembly process. The B-RWK, however, needs to have further work performed to upgrade it to a reusable state (see Figure C-9). The gross requirements for B-REC, reconditioned brush tubes, come from the planned orders for A-REM, remanufactured brush assemblies. Our rework bill of material (see Figure C-4) calls for a 20 percent use of new belt link rings (item X). The scheduled receipts of the item B-RWK brush tubes for rework came from teardown planning (see Figure C-8).

In the example, the projected receipts of tubes for rework (item B-RWK), plus the current inventory, satisfy all requirements. If the supply was less than demand, the system would create an exception message for planner review. The planner has only three choices: Schedule more teardown, which will provide additional inventories of other yielded material and may not be possible because of the number of units available for teardown; increase the supply of new brush tubes; or schedule fewer remanufactured brush assemblies.

The planning for the remanufacturing process is accomplished just like any new-build process. A schedule of 100 remanufactured brush assemblies (item A-REM) would determine the lower-level projected gross requirements based on the quantities per in the reassembly bill of material (see Figures C-6 and C-10).

Each of the components of the parent will have a material plan. The scheduled receipts for reconditioned items reflect the anticipated yield from teardown and rework processes (see Figure C-11). In this example, the teardown yield has been less than planned, and the projected available balance of the C-REC is negative beginning in period 2. Because of the supply of new shafts, production is not impacted. Safety stock on the new material is usually used to buffer against the fluctuations of yield performance. The requirements for new material supply would also

Figure C-9
Rework Plan

Item: B-REC—Reconditioned Brush Tube
Lt.: 1
Inv.: 80

	WEEKS							
	1	2	3	4	5	6	7	8
Proj. Gross Reqts.	50	50	50	50	50	50	50	50
Sched. Receipts								
Proj. Avail. Bal.	30	80	30	80	30	80	30	80
Planned Order Rel.		100		100		100		100

Item: B-RWK—Brush Tubes for Rework
Lt.: 1
Inv.: 170

	1	2	3	4	5	6	7	8
Proj. Gross Reqts.	100		100		100		100	
Sched. Receipts			50	50	50	50	50	50
Proj. Avail. Bal. 170	70	70	20	70	20	70	20	70
Planned Order Rel.								

Item.: X—New Belt Link Rings
Lt.: 6
Inv.: 50

	1	2	3	4	5	6	7	8
Proj. Gross Reqts.	20		20		20		20	
Sched. Receipts					100			
Proj. Avail. Bal. 50	30	30	10	10	90	90	70	70
Planned Order Rel.								

Figure C-10
Remanufacturing Planning

Master Schedule: A-REM
Lt.: 1
Inv. 250
S.S. 100

				WEEKS				
	1	2	3	4	5	6	7	8
Forecast	0	10	60	100	100	100	100	100
Orders	110	80	40	0	0	0	0	0
Total Demand	110	90	100	100	100	100	100	100
Proj. Avail. Bal. 250	140	50						
MPS.			100	100	100	100	100	100

Lower-Level Projected Gross Requirements:

ITEM	QTY. PER	1	2	3	4	5	6	7	8
B-REC	0.5	50	50	50	50	50	50	50	50
B	0.5	50	50	50	50	50	50	50	50
C-REC	0.9	90	90	90	90	90	90	90	90
C	0.1	10	10	10	10	10	10	10	10
D-REC	1.6	160	160	160	160	160	160	160	160
D	0.4	40	40	40	40	40	40	40	40
E	6.0	600	600	600	600	600	600	600	600

reschedule out if the teardown yield was higher than planned, since new material inventory would not decrease as fast as anticipated.

Typically, the same planner is responsible for both new and reconditioned part planning. Negative projected balances of the reconditioned material can be adjusted periodically by adjusting demand.

Another possible way of dealing with this problem is to structure the new part into the reconditioned part with zero lead time and discrete order policy. In this case, any demand that is not satisfied by the reconditioned part creates a projected gross requirement for the new part. (This method is explained in detail in Appendix D.)

YIELD TRACKING

In a remanufacturing environment, the planning percentages are often changing. In the beginning, these percentages are usually *best guess* estimates from engineering. Sometimes, field data on part replacement are helpful in determining planning estimates, but in reality, this planning information typically comes down to someone's judgment with little, if any, hard data. This means that the planning percentages used to forecast teardown yield need to be constantly monitored and updated to represent the actual yields achieved in the remanufacturing operation.

Some companies do this through the material planning organization. Planners track new material exception messages. Constant reorder messages indicate that yields are more or less than planned, which then triggers a yield percentage audit.

In other companies, the total units through teardown are tracked, and an expected yield is calculated for each recovered item. Actual results are compared with the plan, and significant variances or continuing trends trigger an audit. The difficulty with the latter process is that it does require capturing the recovered quantities. In some companies, the

Figure C-11
Remanufacturing Component Planning

Item: C
Lt.: 6
Order Policy: 40
Inv.: 46
S.S.: 20

	WEEKS							
	1	2	3	4	5	6	7	8
Proj. Gross Reqts.	10	10	10	10	10	10	10	10
Sched. Receipts			40					
Proj. Avail. Bal.	36	26	56	46	36	26	56	46
Plnd. Order Rel.							40	

Item: C-Rec
Lt.: 2
Order Policy: Discrete
Inv.: 170
S.S.: 0

	1	2	3	4	5	6	7	8
Proj. Gross. Reqts.	90	90	90	90	90	90	90	90
Sched. Receipts			*90	90	90	90	90	90
Proj. Avail. Bal.	80	−10	−10	−10	−10	−10	−10	−10
Plnd. Order Rel.								

* Reschedule-In Message

recovered material requires reprocessing and is not available as usable material until sometime after the teardown is completed. This timing factor needs to be considered in the actual-to-planned yield results. Some companies track teardown yield to work orders, which achieves an actual quantity variance, but this method requires a level of control and cost not typical in most remanufacturing operations.

Companies that do track yield performance usually utilize some off-line manual or PC capability rather than attempting to modify the core MRP system logic. Also, most companies will investigate the actual-to-planned variances rather than simply make a percentage adjustment. Often the percentage is correct, but the manufacturing process or sorting/inspection procedures have drifted from the original intent, or an unusual batch of product has been cycled through the remanufacturing process.

It is important to realize that the actual yield-to-planning percentage will seldom be the same. Safety stocks of new material will always be a way of life in the remanufacturing world. The amount of safety stock will be controlled partially by the remanufacturer's ability to track and manage yield changes.

KEEPING IT SIMPLE

The system we have described to serve remanufacturing operations assumes that the end items being remanufactured are fairly complex. It also assumes that the operations are fairly job-shop oriented. In many cases, remanufacturing can be more flow oriented, with a great deal less structure and a lot less control. In such environments, capacity planning, shop floor control, and work-in-process accounting are usually unnecessary and highly wasteful, so much simpler systems are applicable. Material planning can be done with a single-level ratio bill of material identifying new parts only. The authors recommend that the teardown, rework, and reassembly operations be designed to maximize flow and minimize wasteful scheduling, inventory, and accounting expense. Where this is not possible, a remanufacturing site can gain and maintain control of its operations using fully structured bills of material, routings, and material and capacity requirements planning. Regardless of the manufacturing environment, remanufacturing companies can expect to achieve the same class A levels of success achievable in new-build facilities.

NEGATIVE QUANTITY PER APPROACH TO REMANUFACTURING

Some companies deal with remanufacturing by setting up bills of material using negative quantity pers for components to allow material requirements planning (MRP) to calculate recovery or teardown yield and to net that yield against requirements for new component supply. One example of this might be a parent with three components, whose component supply is from new and remanufactured materials. In this situation, two bills of material are required. One is for the teardown or recovery process; the other is for the new or rebuild process (see Figure C-12).

The new-build bill of material is fairly straightforward. One component B, two components C, and one component D are required to manufacture or remanufacture a parent A. Teardown item A-TDN is represented in a yield bill of material, so that when it is planned or exploded by MRP, it will create negative gross requirements as the quantity pers are negative. A-TDN yields component B 100 percent of the time and yields component C 75 percent (as two per, this is equal to 1.5 in decimal format). Component D is scrapped.

If we loaded a new-build plan of 100 parent A per period and a teardown plan of 50 parent A-TDN per period, MRP would calculate the

Figure C-12
Negative Quantity Per Bill of Material

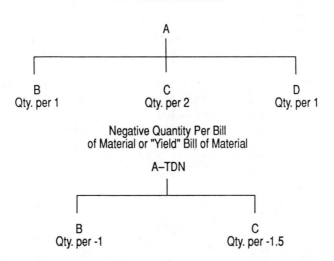

requirements indicated in Figure C-13. The negative and positive requirements would be added to create a *net* gross requirement. The source of the demand would be captured in the MRP pegging logic, and the planner would be able to see the negative and positive demand streams.

If the teardown and new-build demand streams were equal, there would be zero demand for component B, since it has a 100 percent yield planned. If the teardown schedule were greater than the new-build schedule, a negative gross requirement would occur for component B, indicating that the supply of remanufactured components was greater than demand, and an increasing inventory would be projected.

One problem with this approach is lead time offset. In our example, it is assumed that the recovered items flow directly to the new-build process with no lead time consideration. The problem is that recovered components often require some form of remanufacturing before they can be used in parent build and therefore are not available until later periods. If the master schedule is not used for teardown planning, it can be offset to match material availability to rebuild needs.

Another set of problems stems from the fact that negative quantity pers and negative gross requirements give some manufacturing folks real heartburn. In a remanufacturing operation, the people issues are often readily resolved through education.

Figure C-13
Negative Quantity Per Bill of Material Planning

Master Schedule	1	2	3	4	5	6	7	8
A	100	100	100	100	100	100	100	100
A-TDN	50	50	50	50	50	50	50	50

MRP Calculation		
A		B @ 1 PER — 100 × 1 = 100
		C @ 2 PER — 100 × 2 = 200
		D @ 1 PER — 100 × 1 = 100
A-TDN		B @ −1 PER — 50 × −1 = − 50
		C @ −1.5 PER — 50 × −1.5 = − 75

Gross Requirement for Components	1	2	3	4	5	6	7	8
B	50	50	50	50	50	50	50	50
C	125	125	125	125	125	125	125	125
D	100	100	100	100	100	100	100	100

Appendix D

Reprocessing

A problem often encountered by process, pharmaceutical, and other companies is how to structure the bill of material to support reprocessing. To illustrate a common approach to the problem, let's look at an example for manufacturing product Z—a bulk chemical material made from compounds A, B, and C (see Figure D-1).

There are times when ingredient or process differences produce a bulk chemical that does not quite meet specifications. This material is often not scrapped per se, but rather is reprocessed in a future batch of the bulk chemical. A crucial decision in this area is determining how much of this out-of-specification material can be mixed with virgin raw material while still producing the desired product. In our example, the quantity is 30 percent, and the structure for product Z, which reflects reprocessing, is shown in Figure D-2.

Figure D-1
Bill of Material Without Reprocessing

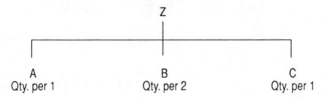

235

Figure D-2
Bill of Material with Reprocessing

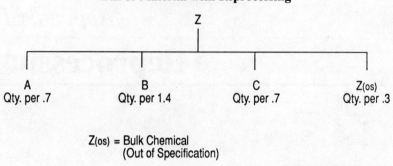

$Z_{(os)}$ = Bulk Chemical
(Out of Specification)

In actual use, a bill of material structure is designed to support the making of bulk chemical product that uses 0 to 30 percent of out-of-specification bulk chemical. This is a two-level bill structure using $Z_{(os)}$ like a phantom that can be stocked (see Figure D-3).

When no out-of-specification bulk chemical stock exists, the plans for manufacturing bulk chemical will blow through the phantom and use all virgin material. Any out-of-specification bulk chemical stock will be used in batches, up to a maximum of 30 percent.

Figure D-4 demonstrates material requirements plan when no stock exists of the out-of-specification bulk chemical. The phantom logic of $Z_{(os)}$ with zero lead time and an order quantity of lot for lot means that the requirements for $Z_{(os)}$ blow through to raw material, of which A is an example.

Figure D-3
Out-of-Specification Item as a Phantom

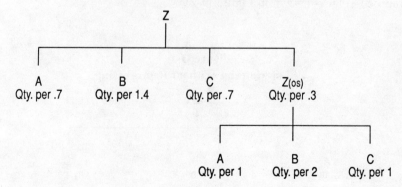

Figure D-4
No Out-of-Specification Material Exists

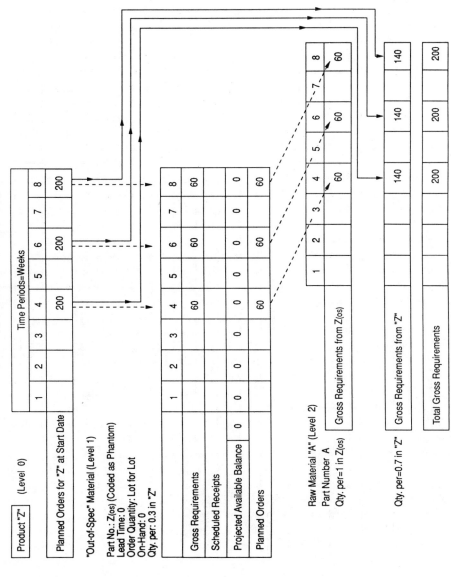

Figure D-5
100 Units of Out-of-Specification Material Exists

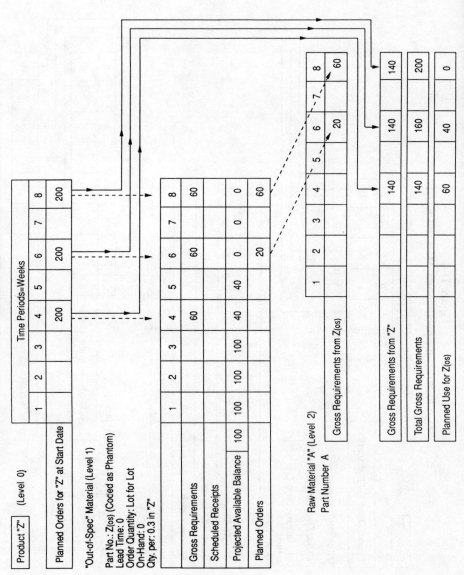

The total requirements for A are 200 in periods 4, 6, and 8. They are made up in those periods from 60 phantoms and 140 bulk chemical main product.

Figure D-5 illustrates the material planning when 100 units of stock exist for the out-of-specification chemical. In period 4, the requirements for 60 $Z_{(os)}$ are met from stock. The maximum amount of reprocessed material, 30 percent, is used. The remaining 70 percent comes from raw materials, of which A is a part.

In period 6, the requirements for 60 $Z_{(os)}$ are met partially from stock, with the remaining 20 blowing through to line up with the A. The reprocessed quantity for this batch is 40 of the 200 kilograms, or 20 percent of the batch.

In period 8, the requirements for 60 $Z_{(os)}$ are entirely blown through to A. In this case, the reprocessed quantity is 0 percent. All the material in this batch is made from virgin raw materials.

Appendix E

Tool Requirements Planning

Tooling, available and in good condition when needed, is a critical issue for many companies. Many adapt their standard material requirements planning and capacity requirements planning systems for this purpose. Both consumable and nonconsumable tools can be planned and scheduled using MRP II capabilities. Consumable tools wear in use, and their replacement, repair, sharpening, or realignment need to be planned. Nonconsumable tools usually don't wear out. Rather, they are simply issued, used, and returned for further use. We want to plan capacity and availability for significant tooling of this type.

Consumable tooling is usually planned using bill of material structures. The tools used in the production of the parent item are given item identifiers and are structured as components of the parent item. There are different quantity per approaches.

In one approach, the quantity per is set at a decimal quantity equal to the tool life consumed in one use of the tool; e.g., $1/1000$, or 0.001, would indicate a tool that can be used 1,000 times. In another approach, the quantity per is set at 1 (see Figure E-1).

Figure E-1
Consumable Tool Components

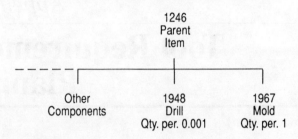

If an order for 200 pieces were planned for the #1246 parent, the dependent demand material planning system would create a demand for 0.2 #1948 drills and for 200 #1967 molds:

COMPONENT	ORDER QTY.		QTY. PER		REQUIREMENTS
1948	200	×	0.001	=	0.200
1967	200	×	1	=	200

Our on-hand and on-order inventory record depends on the approach used to record the quantity per. If the decimal quantity per is used, a new or reconditioned tool would have an inventory value of 1. A tool with 50 percent of its tool life remaining would have a value of 0.50. In the second approach, where the quantity per is 1, a new or reconditioned tool would have an inventory value of the number of times a tool can be used over its tool life. A tool with 50 percent of its tool life remaining would have half of the whole life number (see Figure E-2).

The inventory of tool life remaining for consumable tools needs to be carefully controlled, since the tool life remaining in a given tool might not be readily apparent. This is usually managed with inventory location control, lot control, or a simple manual capability. When a consumable tool is issued from stores, its on-hand work-in-process inventory balance usually is considered available inventory, so that the remaining tool life can be netted against future orders.

In our example, the tool crib personnel must select from three drills in inventory and from three molds in inventory. If the operating rules say to use tools with low remaining tool life and then initiate repair, the tool with the least life would be issued along with the second partially used tool in this example. This will, however, cause a tool change during the job cycle. If the rule is always to issue the tool with more tool life than required, the tool crib will soon be filled with tools that won't support any job requirement.

Some software includes logic that specifies which inventory location to take material from first. Logic of this nature should be reviewed prior to its use in tool planning and control. The right approach is up to the individual company. Some guidelines will be required. One helpful report is a listing of all tools approaching end of tool life so that repair/refurbish activity can be planned or initiated. Specific tools must be tracked, since the planning system will be evaluating aggregate or total tool life in its planning.

The consumable tool planning methodology can also be used to schedule preventive maintenance when usage, rather than time, determines when maintenance is needed. Calibration of equipment is a frequent example. A maintenance identifier is structured into all parent

Figure E-2
Tool Demand and Inventory

Description: Parent Item
Item Number: 1246
Order Number: 696403
Qty. Required: 200

- 1948 Drill Demands: 0.200
  1948 Inventory:
  Location
  TC 113A Drill 1 (New) = 1.000
  TC 114A Drill 2   = 0.170
  TC 114B Drill 3   = 0.430

- 1967 Mold Demand: 200
  1967 Inventory:
  Location
  TC 207 Mold 1 (New) = 1500
  TC 208 Mold 2   = 120
  TC 210 Mold 3   = 690

items that use the equipment that is to be maintained. The on-hand balance is usually set at the number of uses allowed between planned maintenance activity. As usage increases, MRP would move planned maintenance forward, and as usage decreases, MRP would plan for later scheduling.

NONCONSUMABLE TOOLING

When planning nonconsumable tool requirements, the concern is not tool life but rather the amount of time a particular tool will be required and when that tool will be required. In this case, the routing file is the data foundation used to support tool capacity planning and dispatch. One approach involves defining the tool as a work center and adding special tool operations to routings for items that require the tool. Standard work center capacity reporting then projects tool capacities required. One potential problem with this approach is the need to *overlap* the special tool operation with the production operation. The two actually occur simultaneously. This approach also makes it possible to use the work center dispatch list as a scheduling device for tool operations.

In another approach, seen in Figure E-3, tool #1701 is identified in operation step 10 as a requirement in forming the base of parent item #1370. No tool needs are identified for operation steps 20 or 30. It is important to note that most software packages do not provide tool requirements planning using this second approach. Companies usually write some custom code to accomplish this. Some create a separate, stand-alone tool planning subsystem by copying and modifying standard capacity requirements planning and shop floor control logic and

Figure E-3
Routing with Tooling

Parent: 1370
Description: Parent Item

OP.	DEPT.	W.C.	DESCRIPTION	TOOL	SETUP	RUN TIME
10	Fab.	417	Form Base	1701	2.0	0.002
20	Weld.	62	Attach Handle		2.5	0.03
30	Fab.	65	Deburr. Assy.		0.0	0.1

reports. When approached in this way, an interface is required that provides access to schedules with current planning dates.

Whenever a planned or actual order to make #1370 is created, the capacity requirements planning system will determine the number of hours that tool #1701 is required based on the setup and run time needed to produce the number of units on the order. Normally, using a back-scheduling technique, the system will also project when the hours are required (see Figure E-4).

The example shows that order #70631 for 5,000 pieces of item #1370 will require the use of tool #1701 for two shifts. The tool planning system may also add one shift to move a tool to the point-of-use work center and one shift to return it to the tool crib. The total required

Figure E-4
Tool Capacity Planning

Order No.: 70631 Units: 5000
Item No.: 1370 Due Date: 6/29

$$5000 \times 0.002 \text{ (Run Time)} + 2.3 \text{ (Setup)} = 12.3 \text{ Hours}$$
$$12.3 \text{ Hours} = 2 \text{ Shifts} + 2 \text{ Shifts (Move to/From)} = 4 \text{ Shifts}$$

Capacity Summary Report

TOOL 1701	1	2	3	WEEKS 4	5	6	Etc.
Req'd Shifts	10	13	14	11	7	14	
Avail. Shifts	15	15	12	15	12	15	
Diff.	5	2	(2)	4	5	1	

Capacity Detail Week 3

TOOL	ORDER	ITEM	QTY.	SHIFT
1601	69694	1518	2000	5
1601	70606	1670	18000	6
			Total	11
1701	68701	1920	5000	5
1701	69613	2563	19500	5
1701	70631	1370	5000	4
			Total	14
1988	70601	2001	4000	6

Figure E-5
Dispatch List for Tool Scheduling

Date: 6/3

TOOL	ORDER	MOVE DATE	SHIFT	RETURN DATE	SHIFT	SETUP/ RUN HOURS	DEPT.	W.C.
1701	68701	6/4	1	6/5	2	18.4	Fab.	418
1701	69613	6/5	2	6/6	3	16.5	Fab.	417
1601	69694	6/5	3	6/7	1	19.0	Assy.	910
1988	70601	6/6	2	6/8	1	23.5	Fab.	630
1601	70606	6/6	3	6/8	2	26.0	Assy.	920
1701	70631	6/6	1	6/7	1	12.3	Fab.	417
Etc.								

time for tool #1701 for order #70631 is then four shifts. Based on back-scheduling this order from its order due date, this four-shift demand occurs in week 3.

Week 3 is a four-day week, so 12 available shifts are planned. Unfortunately, there are currently 14 shifts of need for our example tool. Review of the week 3 detail report shows that there are three orders that require tool #1701 in that week, totaling 14 shifts of need. A reschedule, if required, would require discussion with the production department and perhaps production control. Fortunately, there is available tool capacity in the surrounding weeks.

Tooling identification on the routing files may be used to provide the tool crib with a dispatch list for tool scheduling (see Figure E-5). This report is in *move date sequence*—the date the toolroom needs to move the tool to the point-of-use work center. It provides the tool crib with the date and shift a tool needs to be issued, the date and shift a tool would return to the tool crib, and the actual hours the tool is required to run an order. The example indicates that tool #1701 has not been rescheduled and is, in fact, scheduled in two instances to be released in the same shift that the tool is moving back to the tool crib. In this case, production and the tool crib have agreed to expedite the tool handling and not reschedule the order operation completion requirements.

There are a variety of approaches in use for tool requirements planning. Regardless of the one chosen, scheduling the right tooling at the right time can be as important as scheduling other materials and other capacities.

Glossary

The terms have been drawn from three primary sources:

1. American Production and Inventory Control Society (APICS) Dictionary 7th Edition, 1992.
2. APICS Process Industry Thesaurus, 1986.
3. Military Standard (MIL-STD-480B) Configuration Control—Engineering Changes, Deviations and Waivers; 15 July 1988.

Unless otherwise noted, the APICS Dictionary is the source document. The designation # (#2) will appear for terms from the Process Industry Thesaurus, and (#3) for those from MIL-STD-480B.

In certain cases, definitions from two or more sources will be listed.* The authors have provided selected definitions when they felt clarification was needed or helpful. In certain cases, definitions have been excerpted. These are noted.

Accessory—A choice or feature offered to customer for customizing the end product. In many companies this term means that the choice does not have to be specified before shipment but could be added at a later date. In other companies, this choice must be made before shipment.
Additives—Special class of ingredients characterized either by being used in minimal quantities or by being introduced into the processing cycle after the initial stage.
Aggregate Reporting—1) Reporting of process hours in general, allowing the system to assign the actual hours to specific products run during the period based upon standards. 2) Also known as "gang reporting," the reporting of total labor hours.
(#2) Aging—Holding of finished product for a specific period of time until product characteristics change to within stated specifications, e.g. cheeses, scotch. The procedure is used to increase

---

* One definition is given when identical or near-identical definitions appear in two sources.

248 Glossary

the value of inventory either by making it usable (initial curing of product) or by transforming it to a new item (12-year-old scotch from 11-year-old scotch). See: incubation period, quarantine.

Allocation—1) In an MRP system, an allocated item is one for which a picking order has been released to the stockroom but not yet sent out of the stockroom. It is an "uncashed" stockroom requisition. 2) A process used to distribute material in short supply. See: reservation.

(#2) Allocation—An allocation is typically a summary number of all requirements for open schedules/orders. This is comparable to a "soft" hold (rather than the "hard" hold of a reservation) in that there is no particular lot assigned to a particular schedule/order. See: reservation.

(#2) Alternate Feedstock—A backup supply of an item to act either as a substitute or to be used with alternate equipment. See: feedstock.

Alternate Operation—Replacement for a normal step in the manufacturing process.

Alternate Routing—A routing, usually less preferred than the primary routing, but resulting in an identical item. Alternate routings may be maintained in the computer or off-line via manual methods, but the computer software must be able to accept alternate routings for specific jobs.

(#2) Alternate Routing—Another procedure for producing the same end item, involving alternate pieces of equipment, differing processing times, and often, an alternative recipe or formula.

(#2) Assays—Report of physical and chemical properties of sample tested by Quality Assurance. Tied by time period to a portion of production.

Assemble-to-Order Product—A make-to-order product for which key components (bulk, semi-finished, intermediate, subassembly, fabricated, purchased, packaging, etc.) used in the assembly or finishing process are planned and stocked in anticipation of a customer order. Receipt of an order initiates assembly of the finished product. This is quite useful where a large number of finished products can be assembled from common components. Syn: finish-to-order.

Assembly—A group of subassemblies and/or parts that are put together and constitute a major subdivision for the final product. An assembly may be an end item or a component of a higher level assembly.

Assembly Lead Time—The time that normally elapses between the time a work order is issued to the assembly floor and its completion.

Assembly Parts List—A list of all parts (and subassemblies) that make up a particular assembly, as used in the manufacturing process. Syn: blend formula, mix ticket.

Attachment—A choice or feature offered to customers for customizing the end product. In many companies, this term means that the choice, although not mandatory, must be selected before the final assembly schedule. In other companies, however, the choice need not be made at that time.

Authorized Deviation—Permission for a supplier or the plant to manufacture an item that is not in conformance with the applicable drawings or specifications.

Available Work—Work that is actually in a department ready to be worked on as opposed to scheduled work, which may not yet be physically on hand. Syn: live load.

(#2) Back Calculated Consumption—Deductions made upon receipt of parent. The determination of usage of raw materials by multiplying receipt quantity of parent times standard quantity per in recipe recognizing standard yield factors. See: backflushing.

Backflush—The deduction from inventory records of the component parts used in an assembly or subassembly by exploding the bills of material by the production count of assemblies produced. Syn: explode-to-deduct.

(#2) Backflushing—Deductions of inventory required at standard made upon receipt of the end item.

(#2) Balanced Loading—Scheduling the production lines to accommodate the limiting rate of one piece of equipment where line balancing is not possible or feasible. Must accommodate both previous and subsequent work stations or lines.

Balancing Operations—In repetitive Just-in-Time production, matching actual output cycle times of all operations to the demand of use for parts as required by final assembly and, eventually, as required by the market.

(#2) Base Stock—A raw material supply for multiple end items; e.g., base gray paint, which is primary ingredient for all colors. See: feed stock.

Batch—A quantity scheduled to be produced or in production. For discrete products the batch is planned to be the standard batch quantity, but during production the standard batch quantity may be broken into smaller lots. In nondiscrete products the batch is the quantity that is planned to be produced in a given time period based on a formula or recipe, which often is developed to produce a given number of end items.

(#2) **Batch**—At standard run or lot size. Determined by vessel size, convention, line rates on standard run length.

(#2) **Batch Bills**—A recipe or a formula whose statement of quantity per for all resources relates to the standard batch quantity of the parent. See: standard batch quantity.

(#2) **Batch/Lot Tracing**—Starting with an end item lot number determining all lot numbers of ingredients/materials consumed to produce the lot number in question. See: batch/lot tracking.

(#2) **Batch/Lot Tracking**—Starting with the lot number of an ingredient, determining all lots into which the lot number in question went.

(#2) **Batch/Mix**—A process business that primarily schedules short production runs of products. See: process/flow.

(#2) **Batch Sensitivity Factor**—A multiplier that is used for the rounding rules in determining the number of batches required to produce a given amount of material.

Batch Sheet—In many process industries, a document that combines product and process definition. Syn: material list, picking list, routing.

(#2) **Batch Sheet**—In many process industries, combines the product and process definition by combining a statement of required materials as well as required manufacturing procedures. See: routing.

Bill of Batches—A method of tracking the specific multilevel batch composition of a manufactured item. The bill of batches provides the necessary where-used and where-from relationships required for lot traceability.

Bill of Capacity—Syn: bill of resources.

Bill of Distribution—Distribution network structure.

Bill of Labor—A structured listing of all labor requirements for the fabrication, assembly, and/or testing of a parent item.

(#2) **Bill of Labor**—See: product load profile. A statement of required labor to complete a process. Stated by labor rate or craft and hours. Used in determining manpower needs. Just as a BOMP could state materials or labor requirements, it can also state resources (all or critical). Syn: product load profile, bill of resources, resource profile.

Bill of Material (BOM)—A listing of all the subassemblies, intermediates, parts, and raw materials that go into a parent assembly showing the quantity of each required to make an assembly. It is used in conjunction with the master production schedule to determine the items for which purchase requisitions and production orders must be released. There is a variety of display formats for bills of material, including the single-level bill of material, indented bill of material, modular (planning) bill of material, transient bill of material, matrix bill of material, and costed bill of material, etc. It may also be called "formula," "recipe," "ingredients list" in certain industries.

Bill of Material Processor—A computer program for maintaining and retrieving bill of material information.

Bill of Material Structuring—The process of organizing bills of material to perform specific functions.

Bill of Operations—Syn: routing.

Bill of Resources—A listing of the required capacity and key resources needed to manufacture one unit of a selected item or family. The resource requirements are further defined by a lead-time offset so as to predict the impact of the item/family scheduled on the load of the key resource by time period. Rough-cut capacity planning uses these bills to calculate the approximate capacity requirements of the master production schedule. Resource planning may use a form of this bill to calculate long-range resource requirements from the production plan. Syn: product load profile, bill of capacity, resource profile.

Blend Formula—An assembly parts list in process industries. Syn: assembly parts list, mix ticket.

Blending—The process of physically mixing two or more lots or types of material to produce a homogeneous lot. Blends normally receive new identification and require testing.

Blend Off—Reworking material by introducing a small percentage into another run of the same product.

(#2) **Blend Off**—Reworking off-spec material by introducing a small percentage back into another run of the same product.
Blend Order—An assembly order in process industries. Syn: assembly order.
Block Control—Control of the production process in groups or "blocks" of shop orders for products undergoing the same basic processes.
Blocked Operations—A group of operations identified separately for instructions and documentation but reported as one.
Blow-through—Syn: phantom bill of material.
Bottleneck—A facility, function, department, or resource whose capacity is equal to or less than the demand placed upon it. For example, a bottleneck machine or work center exists where jobs are processed at a slower rate than they are demanded.
Breeder Bill of Material—A bill of material that recognizes and plans for the availability and usage of by-products in the manufacturing process. The breeder bill allows for complete by-product MRP and product/by-product costing.
Budgeted Capacity—The volume/mix of throughput upon which financial budgets were set and overhead/burden absorption rates established.
Build Cycles—The time period between a major set-up and clean-up. It recognizes cyclical scheduling of similar products with minor changes from one product/model to another.
(#2) **Build Cycles**—Products run between major set-up and major clean-up. Cyclical scheduling of similar product with minor changes from one product/model to another.
(#2) **Bulk Issue**—An issue of material from a controlled stockroom for use on multiple schedules as needed. Issued in quantities more closely aligned to packaging or storage quantities than the planned required quantity for any or all schedules.
By-product—A material of value produced as residual of or incidental to the production process. The ratio of by-product to primary product is usually predictable. By-products may be recycled, sold as is, or used for other purposes.
(#2) **By-product**—An end item incidental to but inevitable from the actual manufacturing process. Not the intended product from a process with minimal potential revenue to the company. A by-product can be garnered from any step of the manufacturing cycle, can be sold as an end item, recycled, or used as raw material for another process. See: co-products, waste.
CAD—Acronym for computer-aided design.
CAD/CAM—The integration of computer-aided design and computer-aided manufacturing to achieve automation from design through manufacturing.
Calculated Usage—The determination of usage of components or ingredients in a manufacturing process by multiplying the receipt quantity of a parent by the quantity of each component/ingredient in the bill/recipe, accommodating standard yields.
Calibration—Maintenance work performed on a tool in order to bring its performance up to an acceptable level or to ensure that current performance levels will be sustained.
Calibration Frequency—Interval in days between tooling calibrations.
CAM—Acronym for computer-aided manufacturing.
(#2) **Campaign**—Sometimes used as a synonym of a schedule or production run of extended duration. Or may be a uniquely important run that requires special attention from all concerned.
Capacity—1) The capability of a system to perform its expected function. 2) The capability of a worker, machine, work center, plant, or organization to produce output per time period. Capacity required represents the system capability needed to make a given product mix (assuming technology, product specification, etc.). As a planning function, both capacity available and capacity required can be measured in the short (capacity requirements planning), intermediate (rough-cut capacity plan), and long term (resource plan). Capacity control is the execution through the I/O control report of the short-term plan. Capacity can be classified as theoretical, rated, demonstrated, protective, productive, dedicated, budgeted, or standing.
Capacity Control—The process of measuring production output and comparing it with the capacity requirements plan, determining if the variance exceeds pre-established limits, and taking corrective action to get back on plan if the limits are exceeded. Syn: input/output control.
Capacity Management—The function of establishing, measuring, monitoring, and adjusting limits or levels of capacity in order to execute all manufacturing schedules, i.e., the production

plan, master production schedule, material requirements plan, and dispatch list. Capacity management is executed at four levels: requirements planning, resource planning, rough-cut capacity planning, capacity requirements planning and input/output control.

Capacity Requirements Planning (CRP)—The function of establishing, measuring, and adjusting limits or levels of capacity. The term "capacity requirements planning" in this context is the process of determining in detail how much labor and machine resources are required to accomplish the tasks of production. Open shop orders and planned orders in the MRP system are input to CRP, which "translates" these orders into hours of work by work center by time period. Syn: capacity planning.

Capacity Simulation—The ability to do rough-cut capacity planning using a "simulated" master production schedule or material plan rather than live data.

(#2) **Carrier Solvent**—Material not consumed during production, but used to transport consumed resources through the process. May also be used to dilute other materials in order to control a reaction.

(#2) **Cascading Yield Loss**—The condition where yield loss happens in sequential operations or tasks resulting in a compounded yield loss.

Cellular Manufacturing—A manufacturing process that produces families of parts within a single line or cell of machines operated by machinists who work only within the line or cell.

Changeover—The refitting of equipment to neutralize the effects of the just-completed production, or to prepare the equipment for production of the next scheduled item, or both. Syn: setup, turnaround, turnaround time.

Changeover Cost—The sum of the teardown costs and the setup costs for a manufacturing operation. Syn: setup cost, turnaround cost.

Charge—The initial loading of ingredients/raw materials into a processor, such as a reactor, to begin the manufacturing process.

(#2) **Chemical Reaction**—An interaction of two or more chemicals resulting in a substance different from the original ingredients.

(#3) **Class I Engineering Change (Excerpt)**—A proposed engineering change shall be determined to be Class I if:
a. Any of the following requirements would be outside specified limits or specified tolerances: (1) performance; (2) reliability, maintainability, or survivability; (3) weight, balance, moment of inertia; (4) interface characteristics; (5) electromagnetic characteristics; or (6) other technical requirements specified in the FCI or ACI.
b. A change impacts one or more of the following:
(1) GFE (Government Furnished Equipment); (2) safety; (3) deliverable operational, test, or maintenance computer software associated; (4) compatibility or specified interoperability; (5) configuration to the extent that retrofit action is required; (6) delivered operation and maintenance manuals; (7) preset adjustments or schedules affecting operating limits or performance; (8) interchangeability, substitutability, or replaceability as applied to CIs; (9) sources of CIs or reparable items at any level; (10) skills, manning, training, biomedical factors, or human engineering design.
c. Any of the following contractual factors:
(1) cost to the Government including incentives and fees; (2) contract guarantees or warranties; (3) contractual deliveries; or (4) schedules contract milestones.

(#3) **Class II Engineering Changes (Excerpt)**—A Class II engineering change identifies a major change that can be affected entirely by a contractor within the scope of a current contract effort. Class II changes are applicable during production and are normally identified and processed to correct documentation errors or to enhance contractor productivity. Examples of Class II engineering changes are:
a. Changes that do not affect interchangeability, substitutability, or replaceability.
b. Substitution of parts or material that do not have a functional, logistic, or reliability impact and
c. Changes in documentation only (e.g., correction of errors, addition of clarifying notes or views, addition, deletion, or correction of non-executable comment lines-of-code to software).

(#2) **Cleanup**—The neutralizing of the effects of production just completed. May involve cleaning of residues, sanitation, equipment refixturing, etc. See: changeover.

Glossary

Closed-loop MRP—A system built around material requirements planning that includes the additional planning functions of sales and operations (production planning, master production scheduling, and capacity requirements planning). Once this planning phase is complete and the plans have been accepted as realistic and attainable, the execution functions come into play. These include the manufacturing control functions of input-output (capacity) measurement, detailed scheduling and dispatching, as well as anticipated delay reports from both the plant and suppliers, supplier scheduling, etc. The term "closed loop" implies that not only is each of these elements included in the overall system but also that feedback is provided by the execution functions so that the planning can be kept valid at all times.

Commonality—A condition wherein given raw materials/ingredients are used in multiple parents.

Common Parts Bill (of Material)—A type of planning bill that groups common components for a product or family of products into one bill of material, structured to a pseudo parent item number.

Component—Raw material, ingredient, part, or subassembly that goes into a higher level assembly, compound, or other item. This term may also include packaging materials for finished items. Syn: intermediate.

Composite Lead Time—Syn: cumulative lead time.

Composite Manufacturing Lead Time—Syn: cumulative manufacturing lead time.

Composite Part—A part that represents operations common to a family or group of parts controlled by group technology. Tools, jigs, and dies are used for the composite part and therefore any parts of that family can be processed with the same operations and tooling. The goal here is to reduce setup costs.

(#2) Composition—The makeup of an intermediate ingredient or finished item, typically expressing chemical properties rather than physical properties. See; specification, assays.

(#2) Compound Yield—The resulting effect of yield loss at multiple operations within the manufacturing cycle. See: cascading yield loss.

Computer-Aided Design (CAD)—The use of computers in interactive engineering drawing and storage of designs. Programs complete the layout, geometric transformations, projections, rotations, magnifications, and interval (cross-section) views of a part and its relationship with other parts.

Computer-Integrated Manufacturing (CIM)—The integration of the total manufacturing organization through the use of computer systems and managerial philosophies that improve the organization's effectiveness; the application of a computer to bridge various computerized systems and connect them into a coherent, integrated whole. For example, budgets, CAD/CAM, process controls, group technology systems, MRP II financial reporting systems, etc., are linked and interfaced.

Computer Numerical Control (CNC)—A technique in which a machine tool control uses a minicomputer to store numerical instructions.

(#2) Concentration—The percentage of active ingredient within the whole, i.e., a 40% solution of HCL—hydrochloric acid. See: potency.

(#3) Configuration—The functional and physical characteristics of hardware, firmware, software, or a combination thereof as set forth in technical documentation and achieved in a product.

Configuration Control—The function of ensuring that the product being built and shipped corresponds to the product that was designed and ordered. This means that the correct features, customer options, and engineering changes have been incorporated and documented.

(#3) Configuration Control—The systematic proposal, justification, evaluation, coordination, approval or disapproval of proposed changes, and the implementation of all approved changes in the configuration of a CI after formal establishment of its baseline.

(#3) Configuration Item (CI) (Excerpt)—An aggregation of hardware, firmware, software, or any of its discrete portions, which satisfies an end use function and is designated for configuration management. CIs may vary widely in complexity, size, and types, from an aircraft, ship, or electronic system to a test meter or round of ammunition.

Continuous Flow (Production)—Lotless production in which products flow continuously rather than being divided.

Continuous Process Control—The use of transducers (sensors) to monitor a process and make automatic changes in operations through the design of appropriate feedback control loops. Although such devices have historically been mechanical or electromechanical, there is now widespread use of microcomputers and centralized control.

(#2) **Continuous Process Run**—A campaign of extended duration. The production is done on dedicated equipment that can produce one product (or product line of slightly varying specifications) without changeover to other products also in demand.

Continuous Production—A production system in which the productive equipment is organized and sequenced according to the steps involved to produce the product. This term denotes that material flow is continuous during the production process. The routing of the jobs is fixed and setups are seldom changed. Syn: mass production.

(#2) **Conversion Costs**—The costs of transforming raw materials (ingredients) into salable product.

Co-products—Products that are usually manufactured together or sequentially because of product and/or process similarities.

(#2) **Co-products**—Similar to by-products except that revenues generated are significant. It may be possible in some instances for the planner to elect to alter the production distribution of individual products in order to balance inventories. See: by-products.

Cost Center—The smallest segment of an organization for which costs are collected and formally reported, typically a department. The criteria in defining cost centers are that the cost be significant and the area of responsibility be clearly defined. A cost center is not necessarily identical to a work center; normally, a cost center encompasses more than one work center, but this may not always be the case.

Costed Bill of Material—A form of bill of material that extends the quantity per of every component in the bill by the cost of the components.

Count Point Backflush—A backflush technique using more than one level of the bill of materials and extending back to the previous points where production was counted. Syn: key point backflush.

Coverter—A manufacturer that changes the products of a basic producer into a variety of industrial and consumer products. An example is a firm that changes steel ingot into bar stock, tubing, or plate. Other converter products are paper, soap, and dyes.

Crew Size—The number of people required to perform an operation. The associated standard time should represent the total time for all crew members to perform the operation, not the net start-to-finish time for the crew.

Critical Path Method (CPM)—A network planning technique for the analysis of a project's completion time used for planning and controlling the activities in a project. By showing each of these activities and their associated times, the "critical path," which identifies those elements that actually constrain the total time for the project, can be determined.

Cumulative Lead Time—The longest planned length of time involved to accomplish the activity in question. For any item planned through MRP, it is found by reviewing the lead time for each bill of material path below the item; whichever path adds up to the greatest number defines cumulative lead time. Syn: aggregate lead time, combined lead time, composite lead time, critical path lead time, and stocked lead time.

Cumulative Manufacturing Lead Time—The cumulative planned lead time when all purchased items are assumed to be in stock. Syn: composite manufacturing lead time.

(#2) **Cumulative Yield**—See: compound yield, cascading yield loss.

(#2) **Cycle Length**—The time between major setups. The time between the start of one production run of similar items/models and the start of a run of the next product/manufacturing family. Syn: cycle time (throughput time). See: build cycles.

Cycle Time—1) In industrial engineering, the time between completion of two discrete units of production. For example, the cycle time of motors assembled at a rate of 120 per hour would be 30 seconds. 2) In materials management, it refers to the length of time from when material enters a production facility until it exits. Syn: throughput time.

(#2) **Cyclical Scheduling**—A method of scheduling product/manufacturing families. A technique to determine run times and quantities for each end item within the family to produce enough of each individual product to satisfy demand until the family can be scheduled again. See: build cycles.

254   Glossary

Dark Factory—A completely automated production facility with no labor.

Data—Any representations such as alphabetic or numeric characters to which meaning can be assigned.

Data Base—A data processing file-management approach designed to establish the independence of computer programs from data files. Redundancy is minimized and data elements can be added to, or deleted from, the file designs without necessitating changes to existing computer programs.

Data File—A collection of related data records organized in a specific manner (e.g., one record for each inventory item showing product code, unit of measure, production costs, transactions, selling price, production lead time, etc.).

Date Code—A label on products with the date of production. In food industries, it is often an integral part of the lot number.

Date Effectivity—A technique used to identify the effective date of a configuration change. A component change is controlled by effective date within the bill of material for the unchanged parent part number.

(#2) Deblend—Where blend off will not result in customer accepted product. The further processing of product to adjust specific physical and chemical properties to within specification ranges. See: blend off.

(#2) Decoupled Schedules—Using surge tanks and/or buffer stocks to pause between tasks. The resultant intermediate may or may not be still considered WIP inventory. See: equivalent units.

Dedicated Capacity—A work center that is designated to produce a single item or a limited number of similar items. Equipment that is dedicated may be special equipment or may be grouped general purpose equipment committed to a composite part.

Dedicated Equipment—Equipment whose use is restricted to (a) specific operation(s) on a limited set of components.

Dedicated Line—A production line "permanently" configured to run well-defined parts, one piece at a time from station to station.

(#3) Defect—Any nonconformance from specified requirements.

Demonstrated Capacity—Proven capacity calculated from actual output performance data, usually expressed as the average number of items produced multiplied by the standard hours per item.

Dependent Demand—Demand that is directly related to or derived from the bill of material structure for other items or end products. Such demands are therefore calculated and need not and should not be forecast. A given inventory item may have both dependent and independent demand at any given time. For example, a part simultaneously may be the component of an assembly and also sold as a service part.

(#3) Deviation—A specific written authorization, granted prior to the manufacture of an item, to depart from a particular performance or design requirement of a specification, drawing, or other document for a specific number of units or a specified period of time. A deviation differs from an engineering change in that an approved engineering change requires corresponding revision of the documentation defining the affected item, whereas a deviation does not contemplate revision of the applicable specification or drawing.

Direct-deduct Inventory Transaction Processing—A method of inventory bookkeeping that decreases the book (computer) inventory of an item as material is issued from stock, and increases the book inventory as material is received into stock by means of individual transactions processed for each item. The key concept here is that the book record is updated coincidentally with the movement of material out of or into stock. As a result, the book record is a representation of what is physically in stock. Syn: discrete issue.

Discrete Manufacturing—Production of distinct items such as automobiles, appliances, or computers.

(#2) Downgrade Profile—A statement of the hierarchy of allowable downgrades. Substitutions of items meeting tighter specifications for those with wider or overlapping specification ranges.

(#2) Downgrading—The substitution of higher grade product for another either in planning or actual fact. May also occur after QA analysis of the actual specifications achieved during production reveals that the product does not fall within "prime product" specification ranges.

Downstream Operation—Task(s) subsequent to the task currently being planned or executed.

Glossary 255

Downtime—Time when a resource is scheduled for operation but is not producing for reasons such as maintenance, repair, or setup.

Earned Hours—A statement of output reflecting the standard hours for actual production reported during the period. Syn: earned volume.

Effective Date—The date on which a component or an operation is to be added or removed from a bill of material or an assembly process. The effective dates are used in the explosion process to create demands for the correct items. Normally, bill of material and routing systems provide for an effectivity "start date" and "stop date," signifying the start or stop of a particular relationship. Effectivity control may also be by serial number rather than date. Syn: effectivity date.

Effectivity—See: effective date.

Efficiency—A measure (as a percentage) of the actual output to the standard output expected. Efficiency measures how well something is performing relative to expectations; it does not measure output relative to any input. Efficiency is the ratio of actual units produced to the standard rate of production expected in a time period, or actual hours produced to standard hours, or actual dollar volume to a standard dollar volume in a time period. For example, if there is a standard of 100 pieces per hour and 780 units are produced in one eight-hour shift, the efficiency is 780/800 multiplied by 100% or 97.5%.

Elemental Parts—Those parts that consist of one piece. No assembly is involved. Syn: piece parts.

End Item—A product sold as a completed item or repair part; any item subject to a customer order or sales forecast. Syn: end product, finished good, finished product.

End Product—See: end item.

Engineering Change—A revision to a blueprint or design released by engineering to modify or correct a part. The request for the change can be from a customer or from production quality control or another department.

(#3) Engineering Change—An alteration in the approved configuration identification of a CI under development, delivered, or to be delivered.

Engineering Drawings—A visual representation of the dimensional characteristics of a part or assembly at some stage of manufacture.

Engineer-to-Order—Products whose customer specifications require unique engineering design or significant customization. Each customer order results in a unique set of part numbers, bills of materials, and routings.

(#2) Equivalent Units—A translation of WIP inventories into equivalent finished goods units. Occasionally exploded back to raw materials for period end valuation of inventories where no WIP inventory account exists in the general ledger.

Exchange Unit—The number of units to be produced before changing the bit, tool, or die.

Fabrication—Manufacturing operations for components as opposed to assembly operations.

Fabrication Level—The lowest production level. The only components at this level are parts (as opposed to assemblies or subassemblies). These parts are either procured from outside sources or fabricated within the manufacturing organization.

Fabricator—A manufacturer that turns the product of a converter into a larger variety of products. For example, a fabricator may turn steel rods into nuts, bolts, and twist drills, or may turn paper into bags and boxes.

Facilities—The physical plant and equipment.

Failsafe Work Methods—Methods of performing operations so that actions that are incorrect cannot be completed. For example, a part without holes in the proper place cannot be removed from a jig, or a computer system will reject "invalid" numbers or require double entry of transaction quantities outside the normal range. Called "poka-yoke" by the Japanese.

(#2) Family—A group of end items whose similarity of design, composition, and manufacture facilitates being planned in aggregate, whose sales performance is monitored together, and occasionally whose cost is aggregated at this level, especially for process products whose differences are minor variations in specifications or specification ranges. See: manufacturing family, manufacturing group.

Feature—An accessory, attachment, or option.

Feature Code—An identifying code assigned to a distinct product feature that may contain one or more specific part number configurations.

Feeder Workstations—An area of manufacture whose products feed a subsequent work area.

256 Glossary

(#2) Feeder Workstations—An area of manufacture whose products are planned to be available for use to feed in a primary work area, often final assembly or fill and packaging. Planning of the primary work area drives the plan for the feeder work station. This plan may be stated as a rate.

(#2) Feed Stock—Material supply for multiple end items; e.g., base gray paint is the primary ingredient (feed stock) of all colors. See: feedstream, base stock.

Feedstream—A supply source for a process.

Final Assembly—1) The highest level assembled product, as it is shipped to customers. 2) The name for the manufacturing department where the product is assembled. Syn: blending department, erection department, pack-out department.

Finished Goods—Syn: end item.

Finished Product—Syn: end item.

Finishing Lead Time—1) The time that is necessary to finish manufacturing a product after receipt of a customer order. 2) The time allowed for completing the product based on the final assembly schedule.

Finish-to-Order—Syn: assemble-to-order.

Firm Planned Order (FPO)—A planned order that can be frozen in quantity and time. The computer is not allowed to automatically change it; this is the responsibility of the planner in charge of the item that is being planned. This technique can aid planners working with MRP systems to respond to material and capacity problems by firming up selected planned orders. Additionally, firm planned orders are the normal method of stating the master production schedule.

(#3) Fit—The ability of an item to physically interface or interconnect with or become an integral part of another item.

Flexible Automation—Short setup times and the ability to switch quickly from one product to another.

Flexible Machine Center (FMC)—An automated system, which usually consists of CNC machines with robots loading and unloading parts conveyed into and through the system. Its purpose is to provide quicker throughput, changeovers, setups, etc. to manufacture multiple products.

Flexible Manufacturing System (FMS)—A group of numerically controlled machine tools interconnected by a central control system. The various machining cells are interconnected via loading and unloading stations by an automated transport system. Operational flexibility is enhanced by the ability to execute all manufacturing tasks on numerous product designs in small quantities and with faster delivery.

Flow Shop—A form of manufacturing organization in which machines and operators handle a standard, usually uninterrupted material flow. The operators generally perform the same operations for each production run. A flow shop is often referred to as a mass production shop, or is said to have a continuous manufacturing layout. The plant layout (arrangement of machines, benches, assembly lines, etc.) is designed to facilitate a product "flow." Some process industries (chemicals, oil, paint, etc.) are extreme examples of flow shops. Each product, though variable in material specifications, uses the same flow pattern through the shop. Production is set at a given rate, and the products are generally manufactured in bulk.

Focused Factory—A plant established to focus the entire manufacturing system on a limited, concise, manageable set of products, technologies, volumes, and markets precisely defined by the company's competitive strategy, its technology, and economics.

(#3) Form—The defined configuration of an item including the geometrically measured configuration, density, and weight, or other visual parameters that uniquely characterize an item, component, or assembly. For software, form denotes the language, language level, and media.

Formula—A statement of ingredient requirements. A formula may also include processing instructions and ingredient sequencing directions. Syn: formulation, recipe.

Formulation—Syn: formula.

(#3) Function—The action or actions that an item is designed to perform.

(#2) Gang Reporting—The reporting of total labor hours to a work center or department.

Gantt Chart—The earliest and best-known type of control chart especially designed to show graphically the relationship between planned performance and actual performance, named after its originator, Henry L. Gantt. It is used for machine loading, where one horizontal line is used to represent capacity and another to represent load against that capacity, or for following job progress where one horizontal line represents the production schedule and another parallel line represents the actual progress of the job against the schedule in time. Syn: job progress chart.

Gapped Schedule—A schedule in which every piece in a lot is finished at one work center before any piece in the lot can be processed at the succeeding work center; the movement of material in complete lots, causing time gaps between the end of one operation and the beginning of the next. It is a result of using a batched schedule at each operation (work center), where process batch and transfer batch are assumed to be the same or equal. Syn: gap phasing, straight line scheduling.

Gateway Work Center—A starting work center.

Generic Processing—A means of developing routings or processes for the manufacture of products through a family relationship, usually accomplished by means of tabular data to establish interrelationships. It is especially prevalent in the manufacture of raw material such as steel, aluminum, or chemicals.

(#2) Graded Products—An item whose specifications of critical chemical or physical properties will differentiate it from another with the same item number. The specification variation may determine its eventual use, cause alterations in other ingredients in formulas for which it is required, and/or alter its worth in the marketplace although not necessarily its processing cost. Graded products may be raw ingredients, intermediates, or finished goods.

(#2) Grades—The sublabeling of items to identify their particular specification make-up and separate this lot from other production lots without changing the item number.

Group Classification Code—A part of material classification technique that provides for designation of characteristics by successively lower order groups of code. Classification may denote function, type of material, size, shape, etc.

Grouping—Matching like operations together and running them together, sequentially, thereby taking advantage of common setup.

Group Technology—An engineering and manufacturing philosophy that identifies the physical similarity of parts (common routing) and establishes their effective production. It provides for rapid retrieval of existing designs and facilitates a cellular layout.

Hard Automation—Use of specialized machines to manufacture and assemble products. Normally, each machine is dedicated to one function, such as milling.

Hardware—1) In manufacturing, relatively standard items such as nuts, bolts, washers, clips, etc. 2) In data processing, the computer and its peripherals.

(#2) Head Box—A storage container of a feedstock. See: hold tanks, surge tank.

(#2) Hold Tanks—A storage container for intermediates, although can also refer to storage vessel for finished goods, raw ingredients, feed stocks, base stocks, etc.

Implode—1) Compression of detailed data in a summary-level record or report. 2) Tracing a usage and/or cost impact from the bottom to the top (end product) of a bill of material using where-used logic.

Implosion—The process of determining the where-used relationship for a given component. Implosion can be single-level (showing only the parents on the next higher level) or multilevel (showing the ultimate, top-level parent). Ant: explosion. Syn: where-used.

(#2) Incubation Period—The length of time required to hold product in order to verify its quality (see: quarantine) or in order to allow a chemical/physical change to happen before further processing; e.g., fermentation.

Indented Bill of Material—A form of multilevel bill of material. It exhibits the highest level parents closest to the left side margin and all the components going into these parents are shown indented to the right of the margin. All subsequent levels of components are indented farther to the right. If a component is used in more than one parent within a given product structure, it will appear more than once, under every assembly in which it is used.

Indented Tracking—The following of all lot numbers of intermediates and ingredients consumed in the manufacture of a given batch of product down through all levels of the formula.

Indented Where-used—A listing of every parent item, and the respective quantities required, as well as each of their respective parent items, continuing until the ultimate end item or level-0 is referenced. Each of these parent items is one that calls for a given component item in a bill of

258   Glossary

material file. The component item is shown closest to the left margin of the listing, with each parent indented to the right, and each of their respective parents indented even further to the right.

Independent Demand—Demand for an item that is unrelated to the demand for other items. Demand for finished goods, parts required for destructive testing, and service parts requirements are examples of independent demand.

(#2) Indirect Measurement—Determining the quantity on hand by (a) measuring the storage vessels and calculating the content's balance quantity, or (b) theoretically calculating consumption of ingredients and deducting them from the on-hand balance.

Ingredient—Syn: component.

(#2) Ingredient—A required material for the manufacture of its parent, specifically material that is purchased as opposed to a processed intermediate. Syn: component.

(#3) Interchangeable Item—An item that:
a. possesses functional and physical characteristics equivalent in performance, reliability, and maintainability to another item of similar or identical purposes; and
b. is capable of being exchanged for the other item without alteration of the item or of adjoining items, except for adjustment or calibration.

Intermediate—Syn: component.

(#2) Intermediates—A semi-processed state that is not usually available for sale to the marketplace. Comparable to a subassembly in the discrete manufacturer by typically held as WIP in the process world often for material handling and storage reasons. Syn: component, subassemblies.

Intermittent Production—A form of manufacturing organization in which the productive resources are organized according to function. The jobs pass through the functional departments in lots and each lot may have a different routing. Syn: job shop.

Internal Setup Time—Time associated with elements of a setup procedure performed while the process or machine is not running. Ant: external setup time.

Interoperation Time—The time between the completion of one operation and the start of the next.

Inventory Shrinkage—Losses of inventory resulting from scrap, deterioration, pilferage, etc.

Islands of Automation—Stand-alone pockets of automation (robots, a CAD/CAM system, numerical control machines) that are not connected into a cohesive system.

Item—Any unique manufactured or purchased part, material, intermediate, subassembly, or product.

Item Master Record—Syn: item record.

Item Number—A number that serves to uniquely identify an item. Syn: part number, product number, stock code.

Item Record—The "master" record for an item. Typically it contains identifying and descriptive data and control values (lead times, lot sizes, etc.) and may contain data on inventory status, requirements, planned orders, and costs. Item records are linked together by bill of material records (or product structure records), thus defining the bill of material. Syn: item master record, part master record, part record.

Job Shop—Syn: intermittent production.

Just-in-Time (JIT)—A philosophy of manufacturing based on planned elimination of all waste and continuous improvement of productivity. It encompasses the successful execution of all manufacturing activities required to produce a final product, from design engineering to delivery and including all stages of conversion from raw material onward. The primary elements of zero inventories are to have only the required inventory when needed; to improve quality to zero defects; to reduce lead times by reducing setup times, queue lengths, and lot sizes; to incrementally revise the operations themselves; and to accomplish these things at minimum cost. In the broad sense it applies to all forms of manufacturing, job shop and process as well as repetitive. Syn: short-cycle manufacturing, stockless production, zero inventories.

(#2) Key Point Backflushing—The theoretical consumption of resources triggered not upon the receipt of the end item but rather triggered through reporting an intermediate quantity produced and passed forward to the next task. The theoretical consumption will consume only resources required for this processing task and all previous processing tasks that are defined as non-reporting, i.e., not also trigger points for key point backflushing.

Lead Time—1) A span of time required to perform a process (or series of operations). 2) In a logistics context, the time between recognition of the need for an order and the receipt of goods. Individual components of lead time can include order preparation time, queue time, move or transportation time, receiving and inspection time. Syn: total lead time.

Lead-time Offset—A technique used in MRP where a planned order receipt in one time period will require the release of that order in an earlier time period based on the lead time for the item. Syn: offsetting.

Level—Every part or assembly in a product structure is assigned a level code signifying the relative level in which that part or assembly is used within the product structure. Normally, the end items are assigned level 0 with the components/subassemblies going into it assigned to level 1 and so on. The MRP explosion process starts from level 0 and proceeds downward one level at a time.

Limiting Operation—The operation with the least capacity in a series of operations with no alternative routings. The capacity of the total system can be no greater than the limiting operation, and as long as this limiting condition exists, the total system can be effectively scheduled by scheduling the limiting operation.

Line—A specific physical space for manufacture of a product that in a flow plant layout is represented by a straight line. This may be in actuality a series of pieces of equipment connected by piping or conveyor systems.

Line Balancing—1) The balancing of the assignment of the elemental tasks of an assembly line to workstations to minimize the number of workstations and to minimize the total amount of idle time at all stations for a given output level. In balancing these tasks, the specified time requirement per unit of product for each task and its sequential relationship with the other tasks must be considered. 2) A technique for determining the product mix that can be run down an assembly line to provide a fairly consistent flow of work through that assembly line at the planned line rate.

Line Loading—The loading of a production line by multiplying the total pieces by the rate per piece for each item to come up with a finished schedule for the line.

(#2) Load Slip—A statement of required materials to fulfill a customer order. Or, a statement of required materials to move to processing for manufacture into an end item. Or, a sub-lot control ticket designating precise production time to the back to specifications determined for a specific sample.

Lost Time Factor—The complement of productivity, that is, one minus the productivity factor. It can also be calculated as the planned hours minus standard hours earned, divided by the planned hours.

Lot Operation Cycle Time—Length of time required from the start of setup to the end of cleanup for a production lot at a given operation, including setup, production, and cleanup.

Lot Size—The amount of a particular item that is ordered from the plant or a supplier. Syn: order quantity.

Lot Traceability—The ability to identify the lot or batch numbers of consumption and/or composition for manufactured, purchased, and shipped items. This is a federal requirement in certain regulated industries.

Low-Level Code—A number that identifies the lowest level in any bill of material at which a particular component may appear. Net requirements for a given component are not calculated until all the gross requirements have been calculated down to that level. Low-level codes are normally calculated and maintained automatically by the computer software. Syn: explosion level.

Major Setup—The equipment setup and related activities required to manufacture a group of items in sequence, exclusive of the setup required for each item in the group.

Make-to-Order Product—A product that is finished after receipt of a customer order. The final product is usually a combination of standard items and items custom designed to meet the special needs of the customer. Frequently long lead-time components are planned prior to the order arriving in order to reduce the delivery time to the customer. Where options or other subassemblies are stocked prior to customer orders arriving, the term "assemble-to-order" is frequently used.

Make-to-Stock Product—A product that is shipped from finished goods, "off the shelf," and therefore is finished prior to a customer order arriving. The master scheduling and final assembly scheduling are conducted at the finished goods level. (Author's note—last statement is true much, but not all, of the time.)

Manufacturing Data Sheet—Syn: routing.

Manufacturing Lead Time—The total time required to manufacture an item, exclusive of lower-level purchasing lead time. Included here are order preparation time, queue time, setup time, run time, move time, inspection time, and put-away time. Syn: manufacturing cycle.

Manufacturing Process—The series of operations performed upon material to convert it from the raw material or semifinished state to a state of further completion and of greater value. Manufacturing processes can be arranged in a process layout, product layout, or cellular manufacturing layout. Manufacturing processes can be planned to support make-to-stock, make-to-order, assemble-to-order, etc., based on the strategic placement of inventories.

Manufacturing Resource Planning (MRP II)—A method for the effective planning of all resources of a manufacturing company. Ideally, it addresses operational planning in units, financial planning in dollars, and has a simulation capability to answer "what if" questions. It is made up of a variety of functions, each linked together: business planning, sales and operations (production planning), master production scheduling, material requirements planning, capacity requirements planning, and the execution support systems for capacity and material. Output from these systems is integrated with financial reports such as the business plan, purchase commitment report, shipping budget, inventory projections in dollars, etc. Manufacturing Resource Planning is a direct outgrowth and extension of closed-loop MRP.

Mass Production—High-quantity production characterized by specialization of equipment and labor. Syn: continuous production.

Master File—A main reference file of information such as the item master file and work center file. Ant: detail file.

Master Production Schedule (MPS)—1) The anticipated build schedule for those items assigned to the master scheduler. The master scheduler maintains this schedule, and in turn, it becomes a set of planning numbers that drives material requirements planning. It represents what the company plans to produce expressed in specific configurations, quantities, and dates. The master production schedule is not a sales forecast that represents a statement of demand. The master production schedule must take into account the forecast, the production plan, and other important considerations such as backlog, availability of material, availability of capacity, management policies and goals, etc. Syn: master schedule. 2) The result of the master scheduling process. The master schedule is a presentation of demand forecast, backlog, the MPS, the projected on-hand inventory, and the available-to-promise quantity.

Master Route Sheet—The authoritative route process sheet from which all other format variations and copies are derived.

(#2) Material Recycle—The removal of material from a later (or last) stage in the manufacturing cycle and re-introducing it into an earlier operation. The material may be a carrier of the active ingredients, or may be unreacted materials re-introduced for further processing.

Material Yield—The ratio of usable material from a given quantity of the same.

Matrix Bill of Material—A chart made up from the bills of material for a number of products in the same or similar families. It is arranged in a matrix with components and parents in rows (or vice versa) so that requirements for common components can be summarized conveniently.

(#2) Metered Issue—A quantity of consumption wherein the determination of actual quantity used was not counted by hand, but rather by meters.

Methods-time Measurement (MTM)—A system of predetermined motion-time standards, a procedure that analyzes and classifies the movements of any operation into certain human motions and assigns to each motion a predetermined time standard determined by the nature of the motion and the conditions under which it was made.

Military Standard—Product standards and specifications for products for U.S. military or defense contractors, units, suppliers, etc.

MIL SPEC—Acronym for military inspection standard.

(#2) Mixing—Blending or stirring.

Mix Ticket—A listing of all the raw materials, ingredients, components, etc. that are required to perform a mixing, blending, or similar operation. This listing is often printed on a paper ticket, which also may be used as a turnaround document to report component quantities actually used, final quantity actually produced, etc. This term is often used in batch processes or chemical industries. Syn: assembly parts list, batch card, blend formula, manufacturing order.

Modular Bill (of Material)—A type of planning bill that is arranged in product modules or options. It is often used in companies where the product has many optional features; e.g., assemble-to-order companies such as automobile manufacturers.

Move Time—The actual time that a job spends in transit from one operation to another in the plant.

Multilevel Bill of Material—A display of all the components directly or indirectly used in a parent, together with the quantity required of every component. If a component is a subassembly, blend, intermediate, etc., all of its components will also be exhibited and all of their components, down to purchased parts and materials.

Multilevel Where-used—A display for a component listing all the parents in which that component is directly used and the next higher level parents into which each of those parents is used, until ultimately all top-level (level 0) parents are listed.

Network Planning—A generic term for techniques that are used to plan complex projects. Two of the best-known network planning techniques are the critical path method and PERT.

(#3) Nonconformance—The failure of a unit or product to conform to specified requirements.

(#2) Non-prime Product—A result of production whose revenue potential is less to the corporation than the product planned for, scheduled, and thought to be produced. See: off-spec product.

Nonscheduled Hours—Hours when a machine is not scheduled for operation; e.g., nights, weekends, holidays, lunch breaks, major repair, and rebuilding.

Nonsignificant Part Numbers—Part numbers that are assigned to each part but do not convey any information about the part. They are identifiers, not descriptors. Ant: significant part numbers.

(#2) Non-stocking Subassembly—A semi-finished state of materials that is held in WIP and may or may not be identified by a user-assigned part number. See: intermediate.

(#2) Offset Quantity—The quantity out of operation X that is required to be produced and moved to the next operation (Y) and whose run time plus move time after the start time of operation X determines the start time of operation Y. Syn: overlap quantity.

(#2) Off-spec Product—A product whose physical or chemical properties fall outside the acceptable range(s). Syn: off-grade.

Operating Efficiency—A ratio (represented as a percentage) of the actual output of a piece of equipment, department, or plant as compared to the planned or standard output.

Operation Description—The details or description of an activity or operation to be performed. This is normally contained in the routing document and could include setup instructions, operating instructions (feeds, speeds, heats, pressure, etc.), and required product specifications and/or tolerances.

Operation Duration—The total time that elapses between the start of the setup of an operation and the completion of the operation.

Operation List—Syn: routing.

Operation Number—A sequential number, usually two, three, or four digits long, such as 010, 020, 030, etc., that indicates the sequence in which operations are to be performed within an item's routing.

Operation Overlapping—Syn: Overlapped schedule.

Operation/Process Yield—The ratio of usable output from a process stage, or operation to the input quantity, usually expressed as a percentage.

Operation Reporting—The recording and reporting of every manufacturing (shop order) operation occurrence on an operation-to-operation basis.

Operation Sheet—Syn: routing.

Operations Sequence—The sequential steps for an item to follow in its flow through the plant. For instance, operation 1: cut bar stock; operation 2: grind bar stock; operation 3: shape; operation 4: polish; operation 5: inspect and send to stock. This information is normally maintained in the routing file.

Option—A choice or feature offered to customers for customizing the end product. In many companies, the term "option" means a mandatory choice—the customer must select from one of the available choices. For example, in ordering a new car, the customer must specify an engine (option) but need not necessarily select an air conditioner (accessory).

Overlapped Schedule—A manufacturing schedule that "overlaps" successive operations. Overlapping occurs when the completed portion of an order at one work center is processed at one or more succeeding work centers before the pieces left behind are finished at the preceding work center(s). Syn: lap phasing, telescoping, operation overlapping. Ant: gapped schedule.

Overlap Quantity—The amount of items that needs to be run and sent ahead to the following operation before the following "overlap" operation can begin. Syn: offset quantity.

Part—Generally, a material item that is used as a component and is not an assembly, subassembly, blend, intermediate, etc.

Part Master Record—Syn: item record.

Part Number—Syn: item number.

Part Record—Syn: item record.

Part Type—A code for a component within a bill of material; e.g., regular, phantom, reference.

Period Capacity—The number of standard hours of work that can be performed at a facility or work center in a given time period.

(#2) Periodic Measurement—A physical inventory or cycle count. More narrowly, an indirect count technique of process products by determining height in a storage vessel and calculating on-hand quantities from that measurement. See: indirect measurement.

PERT—Acronym for program evaluation and review technique.

Phantom Bill of Material—A bill of material coding and structuring technique used primarily for transient (nonstocked) subassemblies. For the transient item, lead time is set to zero and the order quantity to lot-for-lot. This permits MRP logic to drive requirements straight through the phantom item to its components, but the MRP system usually retains its ability to net against any occasional inventories of the item. This technique also facilitates the use of common bills of material for engineering and manufacturing. Syn: blow-through, pseudo bill of material, transient bill of material.

Planning Bill (of Material)—An artificial grouping of items and/or events in bill of material format, used to facilitate master scheduling and/or material planning. It is sometimes called a pseudo bill of material. (Authors note: Term is used by many—with the definition listed under "super bill" in this glossary.)

(#2) Planning Factor—A constant used to multiply against requirements in order to charge the process with enough resources to net out to the requirement quantity after planned yield losses have occurred. See: compound yield.

(#2) Plant Train—A description of a given line or of a product flow within a plant.

Poka-yoke (mistake-proof)—Mistake-proofing techniques, such as manufacturing or setup activity designed in a way to prevent an error from resulting in a product defect. For example, in an assembly operation, if each correct part is not used, a sensing device detects a part was unused and shuts down the operation, thereby preventing the assembler from moving the incomplete part onto the next station or beginning another one. Sometimes spelled "poke-yoke." Syn: failsafe techniques, failsafe work methods, mistake proofing.

Post-deduct Inventory Transaction Processing—A method of inventory bookkeeping where the book (computer) inventory of components is automatically reduced by the computer only after completion of activity on the components' upper level parent item, based on what should have been used as specified in the bill of material and allocation records. This approach has the advantage of a built-in differential between the book record and what is physically in stock. Syn: explode-to-deduct.

Potency—The measurement of active material in a specific lot, normally expressed in terms of an active unit. Typically used for materials as solutions.

(#2) Potency—The concentration of a material, e.g., 52% solution of HCL. See: concentration.

Pre-deduct Inventory Transaction Processing—A method of inventory bookkeeping where the book (computer) inventory of components is reduced prior to issue, at the time a scheduled receipt for their parent or assembly is created via a bill of material explosion. This approach has the disadvantage of a built-in differential between the book record and what is physically in stock.

Predetermined Motion Time—An organized body of information, procedures, techniques, and motion times employed in the study and evaluation of manual work elements. It is useful in categorizing and analyzing all motions into elements whose unit times are computed according to such factors as length, degree of muscle control, precision, etc. The element times provide the basis for calculating a time standard for the operations. Syn: synthetic time standard.

Primary Location—The designation of a certain storage location as the standard, preferred location for an item.

Primary Operation—A manufacturing step normally performed as part of a manufacturing part's routing. Ant: alternate operation.

Primary Work Center—The work center wherein an operation on a manufactured part is normally scheduled to be performed. Ant: alternate work center.

Prime Operations—Critical or most significant operations whose production rates must be closely planned.

Process Capability—The basic physical capability of production equipment and procedures to hold dimensions and other characteristics of products within the acceptable bounds for the process. It is not the same as tolerances or specifications required of the produced units.

(#2) **Process Controllers**—Sophisticated, custom-programmed computers designed to monitor the manufacturing cycle during production. Often with the capability to modify conditions (temperature, flow, pressure, etc.) to bring the production back to within prescribed ranges.

(#2) **Process Engineering**—The discipline or function responsible for the product/process definition.

(#2) **Process/Flow**—A manufacturer who produces with minimal interruptions in any one production run or between production runs of products that exhibit process characteristics such as liquids, fibers, powders, gases. Characterized by the difficulty of planning and controlling quantity yield variances. See: batch/mix.

Process Flow Production—A production approach with minimal interruptions in the actual processing in any one production run or between production runs of similar products. Queue time is virtually eliminated by integrating the movement of the product into the actual operation of the resource performing the work.

Process Hours—The time required at any specific operation or task to process product.

(#2) **Process List**—A listing of procedures in the manufacture of product that may or may not also include a statement of material requirements. See: routing.

Process Manufacturing—Production that adds value by mixing, separating, forming, and/or performing chemical reactions. It may be done in either batch or continuous mode.

Process Sheet—Detailed manufacturing instructions issued to the plant. The instructions may include specifications on speeds, feeds, temperatures, tools, fixtures, and machines and sketches of setups and semifinished dimensions.

Process Steps—The operations or stages within the manufacturing cycle required to transform components into intermediates or finished goods.

(#2) **Process Stocks**—Raw ingredients or intermediates available for further processing into marketable products. See: feed stock.

Process Time—The time during which the material is being changed, whether it is a machining operation or an assembly. Process time per price (setup time/lot size) plus run time per price. Syn: residence time.

(#2) **Process Train**—See: plant train.

Product—Any commodity produced for sale.

Product Configuration Catalog—A listing of all upper level configurations contained in an end-item product family. Its application is most useful when there are multiple end-item configurations in the same product family. It is used to provide a transition linkage between the end-item level and a two-level master production schedule. It also provides a correlation between the various units of upper level product definition.

Product Family—A group of products with similar characteristics, often used in sales and operations (production) planning.

Product Grade—The categorization of goods based upon the range of specifications met during the manufacturing process.

Production Control—The function of directing or regulating the movement of goods through the entire manufacturing cycle from the requisitioning of raw material to the delivery of the finished products.

Production Planning—Syn: sales and operations planning.

Production Line—A series of pieces of equipment dedicated to the manufacture of a specific number of products or families.

Productivity—1) An overall measure of the ability to produce a good or a service. It is the actual output of production compared to the actual input of resources. Productivity is a relative measure across time or against common entities. In the production literature, attempts have been made to define total productivity where the effects of labor and capital are combined and divided into the output. One example is a ratio that is calculated by adding the standard hours of labor actually produced plus the standard machine hours actually produced in a given time period divided by the actual hours available for both labor and machines in the time period. 2) In economics, the ratio of output in terms of dollars of sales to an input such as direct labor in terms of the total wages. This is also called a "partial productivity measure."

Product Line—A group of products whose similarity in manufacturing procedures, marketing characteristics, or specifications allows them to be aggregated for planning, marketing, or occasionally, costing.

Product Load Profile—A listing of the required capacity and key resources required to manufacture one unit of a selected item or family. Often used to predict the impact of the item scheduled on the overall schedule and load of the key resources. Rough-cut capacity planning uses these profiles to calculate the approximate capacity requirements of the master production schedule and/or the production plan. Syn: bill of labor, bill of resources.

(#2) Product Sequencing—A natural progression from one product to another within a family so as to minimize setup and cleanup (switch over) costs. See: cyclical/scheduling.

(#2) Product Specification—A statement of acceptable physical and chemical properties or an acceptable range of properties that distinguish one product from another or one product grade from another.

Product Structure—The sequence that components follow during their manufacture into a product. A typical product structure would show raw material converted into fabricated components, components put together to make subassemblies, subassemblies going into assemblies, etc.

Product Structure Record—A computer record defining the relationship of one component to its immediate parent and containing fields for quantity required, engineering effectivity, scrap factor, application selection switches, etc.

Pseudo Bill (of Material)—Syn: phantom bill of material. (Author's note: "Pseudo" is often used to describe the parent item of an artificial grouping or collection of components that is useful for planning but, unlike phantoms, cannot be manufactured. Like phantoms, pseudos are "blow through" items when planning material. Common parts groups are frequently pseudo items.

(#2) Quality Assurance (QA)—The discipline or function of verifying conformance to specification. May also include the responsibility for standard specification.

(#2) Quarantine—(QC-Hold) The setting aside from availability for use or sale of finished product or raw ingredients until all required quality tests have been performed and conformance to specification or regulations certified.

Queue Time—The amount of time a job waits at a work center before setup or work is performed on the job. Queue time is one element of total manufacturing lead time. Increases in queue time result in direct increases to manufacturing lead time and work-in-process inventories.

Rated Capacity—1) The demonstrated capability of a system. Traditionally, capacity is calculated from such data as planned hours, efficiency, and utilization. The rated capacity is equal to hours available × efficiency × utilization. Syn: calculated capacity, nominal capacity. 2) In theory of constraints, rated capacity = hours available × efficiency activation × availability, where activation is a function of scheduled production and availability is a function of uptime. Syn: standing capacity.

(#2) Raw Materials—Purchased materials (ingredients) to which no processing has been done in house. In accounting, reporting of inventory valuations as a sub-class of inventory that may include intermediates.

(#2) Reactor—A special vessel designed and built to contain a reaction.

(#2) Recycle—The re-introduction of partially processed product or carrier solvents out of one operation or task into a previous operation. A recirculation process. See: blend off, carrier solvent.

Remanufacturing—An industrial process in which worn-out products are restored to like-new condition. In contrast, a repaired or rebuilt product normally retains its identity, and only those parts that have failed or are badly worn are replaced or serviced.

(#3) Repair—A procedure that reduces but does not completely eliminate a nonconformance resulting from production, and which has been reviewed and concurred in by the Material Review Board (MRB) and approved for use by the Government. The purpose of repair is to reduce the effect of the nonconformance. Repair is distinguished from rework in that the characteristic after repair still does not completely conform to the applicable specifications, drawings, or contrast requirements.

Repetitive Manufacturing—A form of manufacturing where various items with similar routings are made across the same process whenever production occurs. Products may be made in separate batches or continuously. Production in a repetitive environment is not a function of speed or volume.

(#3) Replacement Item—An item that is replaceable with another item, but which may differ physically from the original item in that the installation of the replacement item requires operations such as drilling, reaming, cutting, filing, shimming, etc., in addition to the normal application and methods of attachment.

Reprocessed Material—Goods that have gone through selective rework or recycle.

Reservation—The process of designating stock for a specific order or schedule. Syn: allocation.

Resource—Anything that adds value to a product or service in its creation, production, and delivery.

Resource Planning—The process of establishing, measuring, and adjusting limits or levels of long-range capacity. Resource planning is normally based on the production plan but may be driven by higher level plans beyond the time horizon for the production plan, i.e., business plan. It addresses those resources that take long periods of time to acquire. Resource planning decisions always require top management approval. Syn: long-range resource planning, long-term planning.

(#2) Reversibility—The returning of the finished product to its original state. Disassembly.

Revision Level—A number or letter representing the number of times a part drawing or specification has been changed.

(#3) Rework—A procedure applied to a nonconformance that will completely eliminate the nonconformance and result in a characteristic that conforms completely to the specifications, drawings, or contract requirements.

Route Sheet—Syn: operation chart, routing.

Routing—A set of information detailing the method of manufacture of a particular item. It includes the operations to be performed, their sequence, the various work centers to be involved, and the standards for setup and run. In some companies, the routing also includes information on tooling, operator skill levels, inspection operations, testing requirements, etc. Syn: bill of operations, instruction sheet, manufacturing data sheet, operation chart, operation list, operation sheet, routing sheet.

Run Sheet—A log-type document used in continuous processes to record raw materials used, quantity produced, in-process testing results, etc. It may serve as an input document for inventory records.

Run Size—Syn: standard batch quantity.

Run Standards—Syn: run time.

Run Time—Operation setup and processing. Syn: run standards.

Sales and Operations Planning (Production Planning)—The function of setting the overall level of manufacturing output production plan and other activities to best satisfy the current planned levels of sales (sales plan and/or forecasts), while meeting general business objectives of profitability, productivity, competitive customer lead times, etc., as expressed in the overall business plan. The sales and production capabilities are compared and a business strategy that includes a production plan, budgets, pro forma financial statements, and supporting plans for materials and work force requirements, etc., is developed. One of its primary purposes is to establish production rates that will achieve management's objective of satisfying customer demand, by maintaining, raising, or lowering inventories or backlogs, while usually attempting to keep the work force relatively stable. Because this plan affects many company functions, it is normally prepared with information from marketing and coordinated with the functions of manufacturing, engineering, finance, materials, etc. Syn: sales and operations planning.

Scheduled Downtime—Planned shutdown of equipment or plant to reform maintenance or to adjust to softening demand.

Scheduled Load—The standard hours of work required by scheduled receipts, i.e., open production orders.

Scrap—Material outside of specifications and of such characteristics that rework is impractical.

(#2) Scrap—Produced material outside acceptable range of material and of such characteristics that rework is impossible or impractical. Not waste, which is an anticipated by-product. Must be used in addition to yield loss in determining good output to input. See: waste.

Scrap Factor—A percentage factor in the product structure used to increase gross requirements to account for anticipated loss within the manufacture of a particular product. Syn: scrap rate.

Scrap Rate—Syn: scrap factor.

Semifinished Goods—Products that have been stored uncompleted awaiting final operations that adapt them to different uses or customer specifications.

Semiprocess Flow—A manufacturing configuration where most jobs go through the same sequence of operations even though production is in job lots.

(#2) Separating—Removing one substance from a blend or solution.

Sequencing—Determining the order in which a manufacturing facility is to process a number of different jobs in order to achieve certain objectives.

(#2) Sequencing—The prioritization of products within a family that is scheduled cyclically. Prioritization is intended to minimize lost time due to cleanup/setup between products.

Service Parts—Those modules, components, and elements that are planned to be used without modification to replace an original part during the performance of maintenance. Syn: repair parts.

Setup—1) The work required to change a specific machine, resource, work center, or line from making the last good piece of unit A to the first good piece of unit B; 2) The refitting of equipment to neutralize the effects; e.g., teardown of the just completed production and preparation of the equipment for production of the next scheduled item. Syn: changeover, turnover.

Setup Lead Time—The time needed to prepare a manufacturing process to start. Setup lead time may include run and inspection time for the first piece.

Setup Time—The time required for a specific machine, resource, work center, or line to convert from the production of the last good piece of lot A to the first good piece of lot B.

(#2) Setup Time—Preparing equipment and tools for the processing of product. For most process companies, tracked separately, from cleanup time.

Shelf Life—The amount of time an item may be held in inventory before it becomes unusable.

(#2) Shelf Life—The expected number of days a product can be kept in storage and still retain acceptable properties within the standard range of specifications.

Shelf Life Control—A technique of physical first in, first out usage aimed at minimizing stock obsolescence.

Shop Planning—The function of coordinating the availability of material handling, material resources, setup, and tooling so that an operation or job can be done on a particular machine. Shop planning is often part of the dispatching function. The term shop planning is sometimes used interchangeably with dispatching although dispatching does not have to necessarily include shop planning. For example, the selection of jobs might be handled by the centralized dispatching function while the actual shop planning might be done by the foreman or a representative.

Shrinkage—Reductions of actual quantities of items in stock, in process, or in transit. The loss may be caused by scrap, theft, deterioration, evaporation, etc.

Shrinkage Factor—A percentage factor in the item master record that compensates for expected loss during the manufacturing cycle either by increasing the gross requirements or by reducing the expected completion quantity of planned and open orders. The shrinkage factor differs from the scrap factor in that the former affects all uses of the part and its components and the scrap factor relates to only one usage. Syn: shrinkage rate.

Shrinkage Rate—Syn: shrinkage factor.

Significant Part Number—A part number that is intended to convey certain information, such as the source of the part, the material in the part, the shape of the part, etc. These usually make part numbers longer. Ant: nonsignificant part numbers.

Single-level Bill of Material—A display of those components that are directly used in a parent item. It shows only the relationships one level down.

Single-level Where-used—Single level where-used for a component lists each parent in which that component is directly used and in what quantity. This information is usually made available through the technique known as implosion.

SKU—Abbreviation for stockkeeping unit.

Specification—A clear, complete, and accurate statement of the technical requirements of a material, an item or a service, and of the procedure to be followed to determine if the requirements are met.

Standard Allowance—The established or accepted amount by which the normal time of an operation is increased within an area, plant, or industry to compensate for the usual amount of fatigue and/or unavoidable delays.

(#2) Standard Batch Quantity (SBQ)—The normal quantity of production. All ingredient quantities required for the production are stated in terms of the SBQ.

(#2) Standardized Ingredient—A raw ingredient that has been pre-processed to bring all specifications within standard ranges prior to introduction to the main process. Used to minimize variability in recipes.

Standard Time—The length of time that should be required to (1) set up a given machine or operation; and (2) run one part/assembly/batch/end product through that operation. This time is used in determining machine requirements and labor requirements. It is also frequently used as a basis for incentive pay systems as a basis of allocating overhead in cost accounting systems. Syn: standard hours.

(#2) Step Function Scheduling—Logic that recognizes run length to be a multiple of the number of batches rather than simply a linear relationship of run time to total production quantity.

Stockkeeping Unit (SKU)—An item at a particular geographic location. For example, one product stocked at the plant and at six different distribution centers would represent seven SKUs.

(#3) Substitute Item—An item that possesses such functional and physical characteristics as to be capable of being exchanged for another only under specified conditions in particular applications, or from a list of approved substitute parts without alteration of the item or of adjoining items.

Summarized Bill of Material—A form of multilevel bill of material that lists all the parts and their quantities required in a given product structure. Unlike the indented bill of materials, it does not list the levels of manufacture and lists a component only once for the total quantity used.

Summarized Where-used—A form of an indented where-used bill of material listing that shows all parents in which a given component is used, the required quantities, and all of the next level parents until level 0 is reached. Unlike the indented where-used, it does not list the levels of manufacture.

Super Bill (of Material)—A type of planning bill, located at the top level in the structure, that ties together various modular bills (and possibly a common parts bill) to define an entire product or product family. The quantity per relationship of super bill to its modules represents the forecasted percentage of demand of each module. The master scheduled quantities of the super bill explode to create requirements for the modules that also are master scheduled. (Author's note: The term "planning bill" is used by many, including the authors, with the above definition.)

Superflush—A technique to relieve all components down to the lowest level using the complete bill of material, based on the count of finished units produced and/or transferred to finished goods inventory.

(#2) Superflush—Theoretical consumption through multiple levels in the recipe or formula. Typically allows for consumption of subassemblies from stock in the discrete world, but may be used to explode through intermediates in the process world, therefore, not expecting on-hand balances.

(#2) Surge Tank—A container to hold output from one process and feed to subsequent process as needed. Used when line balancing is not possible or practical, or only on a contingency basis when downstream equipment is non-operational. See: head box, hold tanks.

(#2) Target Value—The projected value at standard, i.e., projected cost or standard cost for run quantity. May also be a specific characteristic within a specification expressed as a range; e.g., 28–32% solids.

Teardown Time—The time taken to remove a setup from a machine or facility. Teardown is an element of manufacturing lead tie, but it is often allowed for in setup or run time rather than separately.

Theoretical capacity—The maximum output capability, allowing no adjustments for preventive maintenance, unplanned downtime, shut down, etc.

Time Standard—The predetermined times allowed for the performance of a specific job. The standard will often consist of two parts, that for machine setup and that for actual running. The standard can be developed through observation of the actual work (time study), summation of standard micro-motion ties (predetermined or synthetic time standards), or approximation (historical job times).

Tool Calibration Frequency—The recommended length of time between tool calibrations. It is normally expressed in days.

Tool Number—The identification number assigned to reference and control a specific tool.

Total Quality Control—The process of creating and producing the total composite product and service characteristics by marketing, engineering, manufacturing, purchasing, etc., through which the product and service when used will meet the expectations of customers.

Transient Bill of Material—Syn: phantom bill of material.

(#2) Transport Stocks—A carrier material to move solids in solution or slurry or to dilute ingredients to safe levels for reaction. See: carrier solvent.

(#2) Twilight Materials—Produced material between product runs, scheduled cyclically, but the specifications of this material do not qualify it as prime for either the first product or the second.

U-Lines—Production lines shaped like the letter *U*. The shape allows workers to easily perform several different tasks without much walk time. The number of work stations in a U-line are usually determined by line balancing. U-lines promote communication.

Unit of Issue—The standard issue quantity of an item from stores; e.g., pounds each, box of 12, package of 3, case of 144, etc.

Unit of Measure—The unit in which the quantity of an item is managed; e.g., pounds each, box of 12, package of 3, case of 144, etc.

Utilization—1) A measure of how intensively a resource is being used to produce a good or a service. Utilization measures actual time used to available time. Traditionally, utilization is the ratio of direct time charged (run time plus setup time) to the clock time scheduled for the resource. This measure led to distortions in some cases. 2) In theory of constraints, utilization is the ratio of actual time the resource is producing (run time only) to the clock time the resource is scheduled to produce.

(#2) Variable Bills—Same as Except Bills, which are developed to compensate for variable specifications/characteristics of raw materials actually received, or to compensate for substitutions when standard raw materials are unavailable. See: formula.

Wait Time—The time a job remains at a work center after an operation is completed until it is moved to the next operation. It is often expressed as a part of move time.

(#2) Wash-down—Sometimes more specifically a minor cleanup between similar product runs. Sometimes used in reference to the sanitation process of a food plant. See: cleanup.

(#2) Waste—A by-product with negative value. Waste whole disposal is controlled, or a by-product of a process or task with unique characteristics requiring special management control. Has a negative value. Waste production can usually be planned and somewhat controlled. Scrap (off-spec) is typically not planned and may result from the same production run as waste. See: scrap, yield.

Where-used List—A listing of every parent item that calls for a given component, and the respective quantity required, from a bill of material file. Syn: implosion.

(#2) Where-used Tracking—A procedure to determine every instance of use or sale of a specific lot number, including the use of and/or sale of all parent lot numbers. Parallels the logic of where-used tracing for ingredients/components in bills of material.

Glossary 269

(#3) **Work Breakdown Structure (WBS)**—A product-oriented listing, in family tree order, of the hardware, software, services, and other work tasks that completely defines a product or program. The listing results from project engineering during the development and production of a defense material item. A WBS relates the elements of work to be accomplished to each other and to the end product.

Work Center—A specific production facility, consisting of one or more people and/or machines with identical capabilities, that can be considered as one unit for purposes of capacity requirements planning and detailed scheduling. Syn: load center.

Work Center Where-used—A listing of every manufactured item that is routed (primary or secondary) to a given work center, from a routing file.

Workstation—The assigned location where a worker performs the job; it could be a machine or a work bench.

Yield—The ratio of usable output from a process to its input.

(#2) **Yield**—May involve a loss or increase of quantity or may be a variable distribution of quality/grades, or both. See: grades.

(#2) **Yield History**—Actual yield performance information.

(#2) **Yield Loss-Component**—Yield loss is attributable to a single ingredient/component relationship to the end item.

(#2) **Yield Loss-Item**—Yield is calculated as output divided by all inputs.

(#2) **Yield Loss-Routing**—Loss of input may be tied to a specific operation or task. Sequential occurrences of operational yield loss is cascading yield loss.

(#2) **Yield Planning**—The use of standard or average output quantities and/or distributions to determine required run quantities to meet demand.

(#2) **Yield-Quality**—The distribution of grades within the same end item.

(#2) **Yield-Quantity**—The planned or actual output of a process.

(#2) **Yield Variance**—The difference between planned output and actual; (or) the difference between planned distribution output and actual distribution.

Index

ABCD Checklist, 202–3
Accounting, work-in-process (WIP), 9
Activity-based costing systems, 92–3
Actual costing systems, 90–3
Add/delete bills of material, 133–4
Aftermarket parts sales, 221
Allen, Harvey, 3–5, 100–1
Allocation files (See: Requirements files)
Alternative components, 82–4
Amalgamated planning bills of material, 118
American Production and Inventory Control Society, 18n, 37–8
Application (customer) engineering, 154
As-customer-documented coding, 44
As-designed coding, 41–4
As-manufactured coding, 41–4
As-required quantities, 62–3, 80
Assemble- (finish-) to-order, 136, 195
Assembly sequences, 125
Audit assessment I, 193
Audits:
    bills of material, 66–7, 200
    data foundation change implementation, 200
    routings, 68–9, 200

Backflushing, 44, 96–7, 127–8, 129
Back-scheduling, 85–7
Balloon (find) numbers, 24, 125
Batch cards, 36
Bills of activity, 32, 147, 148, 158
Bills of material (See also: Modularized bills of material)
    accuracy, 62–5, 80, 196
    add/delete, 133–4

alternative names, 25
amalgamated planning, 118
architecture, 49, 99, 105, 107, 196
as-customer-documented, 44
as-designed versus as-manufactured coding, 41–4
audits, 66–7, 200
by-products and co-products, 205–15
completeness, 65, 196
custom manufacturing, 154, 155, 157
data base, 108, 137
definition, 24, 34
disassembly, 205, 223
engineering change control, 170, 175–7, 182–3
exploding, 77, 79, 84, 133
flow versus job shops, 38–9
forward planning, 83
generic, 32, 154, 155, 157
item numbers, 23–4
JIT/TOC, 11, 120
levels, 47, 49–59, 120
master, 118, 133
material requirements planning, 74–5, 77, 79, 208
multiplants, 40–1
negative quantity pers, 233–5
nonstandard, 84–5, 93, 96
order-specific, 84, 175–7
parent/component relationships, 24–30, 47
phantom (transient) items, 33–4, 129
planning (super), 33, 115–9, 195, 196
point-of-use information, 96
preventive maintenance, 217
process industries, 94, 96

271

272   Index

Bills of material *(Cont.)*
  pseudo components, 33–4
  ratio (percentage), 32
  remanufacturing, 223, 225–7, 233–4
  representative, 143–4, 148
  reprocessing, 214, 235–9
  rewards for errors found, 67
  teardown process, 224
  units of measure, 63–5
Blindfolded stockkeeper's test, 20
Block change, 172
Blow-through items, 33–4, 121
By-products, bills of material, 93, 205–15

CAD (computer-aided design), 24, 168
Cancel recommendations, 148
Capacity requirements planning (CRP), 9, 84–7
  alternate routings, 87–90
  back-scheduling, 85–7
  definition, 74, 84
  routing, 84
Cellular manufacturing, 11, 56
  item numbers, 18–9
  routings, 39
Comment (text) files, 36
Common parts groups, 113
Components (See also: End items; Parent/component relationships)
  availability, 73–4
  pick lists, 128
  planning, 30
Computer-aided design (CAD), 24, 168
Concurrent (simultaneous) engineering, 151
Conference room audits, 66, 68
Configuration systems, 132
Configuring-to-order, 101
Construction sheets, 36
Continuous improvement, 11–12, 161
Continuous (mass) production, 38
Co-products, bills of material, 93, 205–15
Correll, James G., 87n
Cost/benefit analyses, 193
Costing systems:
  activity-based, 92–3
  actual versus standard, 9, 90–3, 135, 212
  job, 154
Coverage through dates, 171
CRP (See: Capacity requirements planning)
Cumulative lead time:
  modularized bills of material, 104–9, 111
  planning, 75–7
Customer (application) engineering, 154
Customer order configuration, 44–5

Custom manufacturing, 153–8, 195
  bills of material, 154, 155, 157
  pilot start-up, 158
  routings, 154

Data foundation change implementation, 187–203
  ABCD Checklist, 202–3
  composition, 199–200
  defining structuring activity, 201
  education and training, 198–9
  engineering change control, 201–2
  feedback, 191–2, 199–200
  internal education, 190–2
  method, 199–200
  MRP II and JIT/TQC, 192–6, 198, 200
  project organization, 196–8
  remanufacturing planning and control, 222
  software review, 200
  start point, 200–1
  task sizing, 201
Data foundations, 11–14
  accuracy and completeness, 12–14, 45, 69–72, 92, 97, 196, 200
  administration group, 70
  centralized versus decentralized control, 70
  engineering change control, 167–9
  generic, 154
  item (part numbering), 18–22
  item master data files, 22–3, 36, 51, 74, 82, 114, 120, 200
  JIT/TQC, 11, 45
  live versus master files, 36, 84, 88, 90
  multiplant, 40–1
  operations (work-in-process) detail files, 37, 88
  planning scheduling, 73–97
  reporting methods, 44
  requirements, 12–13, 18–45
  restructuring, 51–9
  routing (work center) master files, 34–5, 36
  software, 44–5
  where-used reporting, 30–2, 44, 95
Date effectivity techniques, 118, 170–5
  add and delete dates, 170–1
  block change, 172
  coverage through dates, 171
  serial number, 172–5
Dates, need versus due, 79–80
Delivery lead time, 104–9, 111
Demand:
  files, 84
  long term, 156
  short term, 156
Dependent demand planning, 74, 77, 80
Design for assembly, 149–50, 152, 153

Index   273

Design for manufacturability, 150–3
Detail scheduling, 156
Discrete transactions, 97
Discrete unit production, 128
Drafting, 24
Drawings, parts lists, 23–4, 125
Due dates, 79–80, 85, 147

ECNs (engineering change notices), 162, 164–7
Economic order quantity (EOQ) formulas, 8
ECRs (engineering change requests), 162, 163–4
Edson, Norris W., 87*n*
End items, 105, 107, 114
   finishing schedule (final assembly), 128–30
   item numbers, 121
   master production schedule, 102–3, 108
   packaging and labeling, 119
End-to-end lead times, 75
Engineering change control, 15, 161–85
   data foundation change implementation, 201–2
   data foundation updating, 167–9
   date effectivity technique, 118, 170–5
   emergency requests, 164
   firm planned orders, 175–82
   historical records, 183–5
   implementation techniques, 170–83
   modular bills of material, 136
   performance measurement, 169–70
   policy and procedure, 162–7
   use-up technique, 182–3
Engineering change notices (ECNs), 162, 164–7
Engineering change request distributions, 164
Engineering change requests (ECRs), 162, 163–4
Engineer-to-order, 136, 153, 158
EOQ (economic order quantity) formulas, 8
Exploding the bill of material, 77, 79, 84, 133

Factory documentation, 129–30
Factory staff audits, 69
Family trees, 28, 29
Feeder stocks, 96
FIFO (first-in, first-out), 96
Final assembly (See: Finishing schedule)
Find (balloon) numbers, 24, 125
Finishing schedule (final assembly), 128–30
Finish- (assemble-) to-order, 136, 195
Finite scheduling software, 39–40
Firm planned orders, 84, 175–82, 212
First-in, first-out (FIFO), 96
Fit, 19–20

Flow shop production, 97
   definition, 38, 120
   versus job shop, 37–40, 97
Forecasting (See: Master production schedule)
Form, 19–20
Formulations, 25
Forward lead time offset, 44
Forward planning (scheduling), 83, 87, 156
Function, 19–20

Gaining Control: Capacity Management and Scheduling (Carrell and Edson), 87*n*
Generic bills of material, 32, 154, 155, 157
Goddard, Walt, 10

Historical records, 183–5

Indented parts list report, 28–30
Interchangeable components, 19, 20, 82–4
Intermediate components, 20, 96, 105, 121
Intermittent production, 37
Inventory:
   files, 36–7
   on hand/on order records, 73
   stocking levels, 20, 119
   tool requirements planning, 243–8
   traditional, 7–9
   work-in-process, 128
Invoicing, 135
Item/location, 40
Item master data files, 51, 74, 114, 120, 200
   accuracy, 69–70
   definition, 22–3
   shrinkage factor, 82
Item (part) numbers, 18–22
   assignment, 18–20
   dummy, 143, 148
   end items, 121
   format, 20–2
   location codes, 40
   maintenance, 217
   refurbished parts, 222

Job costing systems, 154
Job shop production:
   definition, 37
   versus flow shop, 37–40, 97
Just-in-Time/Total Quality Control (JIT/TQC), 9–11
   bills of material, 11, 120
   data foundations, 11, 45, 192–6, 198
   definition, 10
   design for manufacturability, 152, 153
   end items, 108
   item numbers, 18
   waste elimination, 10, 153
   waste identification, 10

274   Index

Just-in-Time: Making It Happen (Sandras), 10
Just-in-Time: Surviving by Breaking Tradition (Goddard), 10

Kanban, 11, 50, 65
  definition, 56
Keys, 125

Labeling, 119
Labor:
  flexibility, 11, 50
  reporting, 92
Lead time, 74–8, 227
  configuring-to-order, 101
  material requirements planning, 74–8
  modularized bill of material, 104–9
  offsets, 121, 125, 128, 132, 134
  phantom (transient) items, 120, 214
Level-by-level time-phased planning, 80
Line sequence numbers, 121, 125, 134–5
Live versus master files, 36, 84, 88, 90
Lots:
  discrete (lot-for-lot) size, 120
  traceability, 95–6, 136

"Make complete," 39
Make parts, 218
Make-to-order, 105, 108, 136, 195
Make-to-stock, 136, 195
Manufacturing instructions, 35
Manufacturing process instructions, 36
Manufacturing Resource Planning (MRP II):
  custom manufacturing, 154, 156, 158
  data foundation change implementation, 192–6, 198, 200
  definition, 7
  preventive maintenance, 217
  product launch, 141–2, 149, 153
  remanufacturing planning, 227
  routing capabilities, 39
  tool requirements planning, 241, 245
Manufacturing restructure, 50–7
Mass (continuous) production, 38
Master bills of material, 118, 133
Master production schedules (MPS), 77
  end items, 102–4, 108
  modularized bills of material, 99, 102
  planning bills of material, 118
  product launch, 142
Master versus live files, 36, 84, 88, 90
Material requirements planning (MRP), 74–82
  bills of material, 74–5, 77, 79, 208
  definition, 73–4
  dependent demand, 74, 77, 80
  engineering change control, 170–2

lead times, 74–8
level-by-level time-phased, 80
master production schedules, 77
planning horizons, 77
Material variance, 92
Melt orders, 36
Menu-driven process, 108, 137
Milliken, Charles, 3–6, 100–1, 104
Mix tickets, 36
Modularized bills of material, 14, 99–137, 196
  add/delete, 133–4
  coding, 121
  construction, 114
  integrating routings, 119–28
  order entry, 130–3
  perspective, 134–6
  planning bills of material, 115–9, 130
  product and business considerations, 102–9
  steps, 136–7
Move date sequence, 248
MPS (See: Master production schedule)
MRP (See: Material requirements planning)
MRP II (See: Manufacturing Resource Planning)
MRP II Standard System, 45
Multiplant manufacturing, 40–1

Need dates, 79–80, 147
Negative gross requirements, 208–11, 235
Negative quantity per approach, 233–5
Network project plans, 32
New product introduction (See: Product launch)
Number (operation) identifiers, 34

"One Less at a Time" approach, 10, 161
One-time bills of material, 84
On hand/on order inventory records, 73
Operation (number) identifiers, 34
Operations (work-in-process) detail files, 37, 88, 90–3
Operation standards, 34
Option percentage, 118
Option sensitivity, 114, 118–9
Order-entry process:
  custom, 130–3
  definition, 128
  menu-driven, 108, 130, 137
  planning bills of material, 130–3, 135, 137
Order-specific bills of material, 84, 175–7
Out-of-balance conditions, 74

Index 275

Packaging, 119
Parent/component relationships, 24–30, 36, 132
 line sequence numbers, 125
 multilevel, 24, 27–30
 single-level, 24–7, 30
Parent-level item audit, 66–7
Parent operation number linkages, 121, 125–8, 135
Pareto analysis, 66
Part numbers (See: Item numbers)
Parts lists, 23–4
Phantom (transient) items, 33–4, 119–21, 129, 135, 214
Planning (super) bills of material, 33, 115–9, 130–3, 195, 196
Planning horizons, 77
Point-of-use information, 96, 121, 123–5, 128, 132, 134
Preventive maintenance planning and control, 217–9
Pricing (See: Costing systems)
Process manufacturing, 93–7
 backflushing, 96–7
 bills of material, 94
 definition, 38
 planning bills of material, 119
 potency or concentration considerations, 96
 product sequencing and campaigns, 94–5
 routing, 94
Process variance, 92
Product family codes, 95
Production-process audits, 68–9
Product launch, 15, 139–59
 design for assembly, 149–50
 design for manufacturability, 150–3
 JIT/TQC, 152, 153
 Manufacturing Resource Planning II, 141–2, 149, 153
 team approach, 142–9
Product liability, 185
Product sequencing, 94–5
Product structure record, 125
Projected gross requirements, 77, 79
Pseudo components, 33–4, 114, 121, 128, 132, 135

Quality review, 221
Queue (wait time), 34, 39, 85

Rapid engineering change control process, 143
Recipe cards, 36
Recipes, 25
Reference quantities, 80

Refurbishing, remanufacturing, and reconditioning, 221–35
 bills of material, 223, 225–7, 233–5
 cost effectiveness, 221–2
 definition, 221
 item numbers, 222
 negative quantity per approach, 233–5
 planning, 227–30
 routings, 224
 yield tracking, 230–2
Remanufacturing (See: Refurbishing, remanufacturing, and reconditioning)
Repetitive manufacturing:
 backflushing, 96–7, 129
 block change, 172
 definition, 38
 process information, 128–9
Replenishment, 74, 79
 kanban, 11, 50, 56, 65
Representative bills of material, 143–4, 148
Reprocessing, 237–41
Requirements files, 37, 84, 90–3
Rescheduling, 118
Rework operations, 87–90
Routings, 14
 accuracy, 67–8, 87
 alternate, 87–90
 audits, 68–9, 200
 capacity requirements planning, 84
 custom manufacturing, 154
 definition, 9, 34
 flow versus job shops, 38, 39
 generic, 154, 155
 integrating modularized bills of material, 119–28
 JIT/TQC, 11
 preventive maintenance, 217–9
 process industries, 94
 product launch, 148
 steps, 47, 50–9
 work center master files, 34–5
Run time field, 218

Safety stock, 218, 232
Sales literature, 135
Sales territory, 118
Same-as-except features, 44
Sandras, Bill, 10
Scheduled receipts, 208, 209, 211, 228
Scrap, 195, 221
Seasonality, 118
Serial number effectivity technique, 172–5
Setup field, 218
Shelf life, 95–6
Shop floor control systems, 156, 225
Shrinkage, 82, 195

Simultaneous (concurrent) engineering, 151
Specifications (bills of material), 25
Specifications (drawings), 23–4
Stacked lead times, 75
Standard costing systems, 91–3, 135
Stocking:
  levels, 20, 119
  strategies, 136
Subassemblies, 121
Subcontractors, 89
Super (planning) bills of material, 33, 115–9, 130–3, 195, 196
Supervisors' audit, 67

Temporary substitute components, 82–4
Text (comment) files, 36
Time-phased allocations files, 84
Tool requirements planning, 241–6
  consumable tools, 241–4
  nonconsumable tools, 241, 244–6
Total Quality Control (See: Just-in-Time/Total Quality Control)
Traceability requirements, 94, 95–6, 136
Transient (phantom) items, 33–4, 119–21, 129, 135, 214
Tying bills of material to orders, 84

Units of measure, dual, 63–5
Universal manufacturing equation, 84
Use-up technique, 182–3

Vision statements, 193

Wait time (queue), 34, 39, 85
Warranties, 136, 221
Waste, 65, 161
  definition, 10
  JIT/TQC, 10, 153
Where-used reporting, 30–2, 44, 95
Work center master files, 200
Work centers, 9
  grouping, 11
  master files, 34–5, 69–70, 74
Work-in-process (WIP) accounting, 9
Work-in-process (WIP) inventory, 128
Work-in-process (operations) detail files, 37, 88, 90–3
Work orders:
  audits, 66
  release logic, 212

Yield, 195, 212, 228–32